FLORA ZAMBESIACA

Flora terrarum Zambesii aquis conjunctarum

T0136670

VOLUME THIRTEEN: PART FOUR

FLORA ZAMBESIACA

MOZAMBIQUE

MALAWI, ZAMBIA, ZIMBABWE

BOTSWANA

VOLUME THIRTEEN: PART FOUR

Edited by
J.R. TIMBERLAKE & E.S. MARTINS

on behalf of the Editorial Board:

D.J. MABBERLEY
Royal Botanic Gardens, Kew

M.A. DINIZ
*Centro de Botânica, Instituto de Investigação
Científica Tropical, Lisboa*

J.R. TIMBERLAKE
Royal Botanic Gardens, Kew

Published by the Royal Botanic Gardens, Kew,
for the Flora Zambesiaca Managing Committee
2010

PLANTS PEOPLE
POSSIBILITIES

First published in 2010 by
Royal Botanic Gardens, Kew
Richmond, Surrey, TW9 3AB, UK
www.org.uk

ISBN 978 1 84246 194 5

British Library Cataloguing in Publication Data
A catalogue record for this book is available from the British Library

Typesetting by Christine Beard
Publishing, Design & Photography
Royal Botanic Gardens, Kew

Printed in the USA by The University of Chicago Press

For information on or to purchase all Kew titles please visit
www.kewbooks.com or e-mail publishing@kew.org

All proceeds go to support Kew's work in saving the world's plants for life

CONTENTS

FAMILIES INCLUDED IN VOLUME 13, PART 4

XYRIDACEAE

ERIOCAULACEAE

TYPHACEAE

RESTIONACEAE

FLAGELLARIACEAE

JUNCACEAE

MUSACEAE

STRELITZIACEAE

COSTACEAE

ZINGIBERACEAE

CANNACEAE

MARANTACEAE

LIST OF NEW NAMES PUBLISHED IN THIS PART

XYRIDACEAE

by J.M. Lock

Annual or perennial herbs, often in nutrient-poor seasonally or permanently wet sites. Stems upright, base sometimes swollen and bulb-like in perennial species. Leaves alternate, simple, linear, in 2 rows, few to numerous; basal sheath open, blade flattened to terete. Inflorescence pedunculate; flowers in axils of densely crowded, often coriaceous bracts, forming a spherical, ovoid or elongate head. Flowers trimerous, bisexual. Calyx lobes 3; the inner at first forming a hood-like structure over the bud, the laterals smaller. Corolla tubular, divided above into 3 broad spreading petals, usually yellow but sometimes white or orange (or blue, but not in Flora area). Stamens 3, opposite petals, often with 3 alternating staminodes. Ovary (1)3-locular, with numerous ovules and axile or parietal placentation; style 1, sometimes divided into 3 at apex. Fruit a loculicidal or irregularly dehiscent capsule. Seeds numerous, small, with endosperm.

A medium-sized pantropical family of mostly small annual or perennial herbs with 5 genera and c.260 species, of which 240 are in *Xyris*. The family is most diverse in South America, and extends into the warm temperate parts of North America; about 35 species in the Flora Zambesiaca area.

In South America the inflorescences of a few species are collected and sold as dried flowers. The only other economic importance is as weeds of wet places, particularly rice fields.

XYRIS L.

Xyris L., Sp. Pl.: 42 (1753); Gen. Pl. ed.5: 25 (1754). —Malme in Bot. Jahrb. Syst. **30**: 287–308 (1912).

Annual or perennial herbs, often forming clumps. Rhizomes when present, short, horizontal or ascending. Leaves in 2 rows, few to many, virtually all basal, flat to terete; differentiated into sheath and lamina, ligule present or not. Inflorescence a condensed bracteate spike at the end of elongated peduncles, which are often variously grooved or ridged. Lower bracts of inflorescence sterile, occasionally elongated or reflexed. Flowers borne singly in axils of fertile bracts, bisexual, trimerous, short-lived. Calyx of 3 unequal segments, the outer hooded and enclosing the flower-bud, the laterals asymmetrical and often keeled, free from each another (sometimes fused basally in American species). Corolla tubular, divided above into 3 broad spreading petals, usually yellow. Stamens 3, opposite petals, alternating with 3 usually bifurcate staminodes divided apically into many moniliform hairs. Ovary (1)3-locular, with numerous ovules and (in the Flora area) parietal placentation; style 1, sometimes divided into 3 at apex. Fruit a loculicidal capsule. Seeds numerous, small, with endosperm.

Useful descriptions and illustrations of character states are given by Kral (Ann. Missouri Bot. Gard. **75**: 522–722, 1988).

All the African species belong to Sect. *Xyris,* characterised by parietal placentation in a unilocular ovary. Of the other two sections, *Pomatoxyris* occurs only in Australia and *Nematopus* only in Central and South America. However, Kral (1988) has cast doubt on the validity of the sections because intermediate types of placentation occur.

Although species in the genus generally look rather different, accurate identification depends on the observation of small features for which a ×10 or ×20 lens or, preferably, a binocular microscope is needed. Anatomical characters may be useful, but have not been investigated for many African species. Characters of value, which should be borne in mind and recorded when collecting, include:

Life-span – annual or perennial; if perennial, is there a swollen bulb-like base? Collect if present.

Leaf sheaths – these should be complete and clean. Surface texture and colour are useful.

Leaf apices – make sure these are present and not grazed or broken off.

Leaf surface ornamentation – spare leaves for sectioning.

Peduncles – spare material for sectioning.

Bracts of inflorescence – a range of ages is useful; include spare inflorescences of various ages for dissection.

Lateral sepals are curved and keeled – most easily removed from a mature or fruiting inflorescence; they are stiff, awkward to handle and easy to lose. The keel may be entire or variously dentate or ciliate. Examine a detached lateral sepal from both sides at ×20 or more to get a clear view of the keel.

Flowers – do not dry well and may open for only part of one day. Spirit material is useful.

Fruits – usually present in mature inflorescences, but tend to dehisce and lose their seeds.

Seed shape and surface ornamentation – seeds must be ripe; use ×20 or ×40 lens.

Many species have minute discontinuous ridges (rugulose) on the leaf and sometimes stem surfaces, visible with a ×20 lens and clearly shown in Fig. 13.4.**2**.

The flowers of some species have been noted as opening in the morning. Doust & Conn (Austral. Syst. Bot. **7**: 455–484, 1994) give opening times for flowers of species they describe; similar data for African species would be welcome.

The family has probably been under-collected in Africa and this account should be regarded as more preliminary than many in the Flora. A summary of the genus in the Flora area was published in 1999 (Lock in Kew Bull. **54**: 301–326). Since then, several new species have been described by the late Prof. S. Lisowski (*Xyris* nouveaux (Xyridaceae) du Haut-Katanga, Syst. Geogr. Pl. **69**: 205–214, 1999), some of which occur within the Flora area.

There is little doubt that new taxa remain to be found and described. Northern and western Zambia and the Chimanimani Mts in Zimbabwe may well support undescribed species. In particular, species with bulbous bases, often flowering precociously, need further study and collection of correlated material at all stages of growth. Likewise, the small annual species, often flowering when less than 10 cm tall, are often overlooked or collected as mixed gatherings. The small African taxa of *Xyris* remain in need of detailed study on a continent-wide basis.

Conservation notes: The assessments given in the following account are based on general impressions from collections seen and not on field experience. They should be seen, however, against four general features. First, almost all the species are plants of seasonally wet places so drainage and cultivation of such sites pose a general risk. Secondly, it appears that most species occur in sites of low nutrient status. While there is no experimental evidence, it is possible that addition of fertilisers or other soil improvement measures may pose a significant risk. Thirdly, none of the species appears to be particularly common anywhere; exactly why this should be is not clear but may be connected with the first two points. Finally, species that occur in shallow soil over rock are vulnerable to the increased use of rock outcrops near settlements for drying clothes, cassava and other materials, as well as to small-scale quarrying for building stone.

1. Outermost inflorescence bracts markedly longer than inner ones ·········2
   - Outermost inflorescence bracts not markedly longer than inner ones ······5
2. Outer inflorescence bracts very pale, almost white, at least at margins ······3
   - Outer inflorescence bracts similar in colour to inner ones, brownish ·······4
3. Outer inflorescence bracts erect, thin, papery and acute at apex, almost white with pale buff central area; leaf surfaces strongly rugulose ······ **1.** *kornasiana*
   - Outer inflorescence bracts spreading, rigid, acuminate; inner bracts dark with bristle-like point··································· **2.** *aristata*
4. Outermost inflorescence bract longer than young inflorescence; Zambia and Angola ·· **9.** *foliolata*
   - Outermost inflorescence bract only slightly elongated; coastal SE Africa ····· ··· **8.** *natalensis*
5. Leaves strongly flattened in one plane, fringed with long flexuous hairs······ ··· **23.** *imitatrix*
   - Leaf margins entire or with small spines ·······························6
6. Keel of lateral sepals entire and glabrous·····························7
   - Keel of lateral sepals toothed or hairy, at least in middle ················16
7. Inflorescence bracts with distinct subterminal triangular or rhomboid mark; inflorescence subspherical to broadly ovoid at maturity··········· **22.** *anceps*
   - Inflorescence bracts without such a mark; inflorescence variously shaped when mature ···8
8. Fertile inflorescence bracts with distinct dorsal ridge··················9
   - Fertile inflorescence bracts smoothly curved dorsally··················10
9. Inflorescence 5–6 mm long, ellipsoid; mainly Mts Mulanje and Namuli ······ ··· **26.** *makuensis*
   - Inflorescence 4–5 mm long, obtriangular; widespread·········· **27.** *huillensis*
10. Leaf sheaths smooth; lamina minutely papillose on margins only; inflorescence subspherical when mature, up to 8 mm in diameter, golden-brown ·········· ··· **29.** *welwitschii*
    - Leaf sheaths smooth, or minutely rugulose over their whole surface·······11
11. Leaf sheaths minutely rugulose, at least towards base·················12
    - Leaf sheaths smooth throughout ·································14
12. Mature inflorescence ellipsoid; bracts yellowish brown; outline of keel smoothly and shallowly convex ······························ **30.** *straminea*[1]
    - Mature inflorescence subspherical; bracts yellowish or reddish··········13
13. Mature inflorescence bracts reddish; sheaths markedly rugulose···· **31.** *rubella*
    - Mature inflorescence bracts yellowish; sheaths weakly rugulose ·· **34.** *fugaciflora*
14. Mature inflorescences at least 5 mm in diameter, usually more ···· **24.** *capensis*
    - Mature inflorescences up to 4 mm in diameter ····················15
15. Inflorescence bracts reddish, spathulate ················· **32.** *schliebenii*
    - Inflorescence bracts brownish, ovate or elliptic················ **33.** *parvula*
16. Inflorescence bracts with distinct subterminal triangular or rhomboid mark ·· 17
    - Inflorescence bracts without such a mark·······················18
17. Annual; inflorescence bracts hairy all over; leaves uniformly coloured when dry ··· **35.** *laniceps*
    - Perennial; inflorescence bracts only hairy towards apex, or glabrous; leaves speckled when dry··································· **21.** *angularis*
18. Inflorescence bracts pale yellow or whitish, hard, shiny ·············19
    - Inflorescence bracts not as above; leaves and sheaths rugulose or not······21

---

[1] Note: Easily confused with *X. huillensis*, which is a small perennial with obtriangular inflorescences and obtuse to rounded apex to lower sterile bracts.

19. Inflorescence bracts pale or straw-yellow; tips of buds whitish or yellowish; mature inflorescence 8–14 mm in diameter · 20
–   Inflorescence bracts whitish; tips of buds wine-red; mature inflorescence 5–8 mm in diameter · **10.** *friesii*
20. Leaves and peduncles rugulose throughout · · · · · · · · · · · · · · · · · · **11.** *lejolyana*
–   Leaves and peduncles smooth · **12.** *symoensii*
21. Fertile bracts reddish brown, papillose towards rounded apex; leaves usually rugulose · **28.** *rhodolepis*
–   Fertile bracts yellowish, blackish or brownish, papillose or not; leaves rugulose or not · 22
22. Fertile inflorescence bracts with distinct dorsal ridge; sterile bracts with thin pale papery margins · **25.** *porcata*
–   Fertile inflorescence bracts without distinct dorsal ridge or, if ridged, then sterile bracts not as above · 23
23. Bracts blackish, hard, opaque, usually acute, acuminate or bristle-tipped · · · 24
–   Bracts blackish or brownish, variously textured, sometimes mucronate but not bristle-tipped· 27
24. Bracts with broad pale scarious margins which become ragged with age · · · · 25
–   Bracts without such margins · 26
25. Bracts markedly papillose, with a long rigid point · · · · · · · · · · · · · · **2.** *aristata*
–   Bracts not markedly papillose, acute but without a long rigid point · · **4.** *gerrardii*
26. Small plants, usually less than 50 cm tall; leaves less than 1 mm wide; inflorescence up to 6 mm long; upper fertile bracts abruptly acuminate or muronate· **5.** *obscura*
–   Robust plants, usually more than 50 cm tall; leaves more than 2 mm wide; inflorescences more than 8 mm long; fertile bracts acute · · · · · · · **3.** *rehmannii*
27. Keel of lateral sepals bearing a row of stellate hairs · · · · · · · · · · · **14.** *asterotricha*
–   Keel of lateral sepals bearing simple hairs or teeth· · · · · · · · · · · · · · · · · · 28
28. Inflorescence sub-spherical; bracts olive-brown, the lower ones spreading; lateral sepals almost parallel-sided, almost the same width throughout their length; leaves and stems smooth, not rugulose· **16.** *peteri*
–   Inflorescence ellipsoid to subspherical; bracts variously coloured, generally appressed; lateral sepals broader towards the apex; leaves and stems rugulose or not· 29
29. Base of stem enlarged to form a bulb-like structure · · · · · · · · · · · · · · · · · 30
–   Base of stem not enlarged · 34
30. Sheaths much expanded towards base, black, shiny; inflorescence bracts dark brown to blackish, shiny, rounded at apex· · · · · · · · · · · · · · · · · · **7.** *gossweileri*
–   Sheaths not as above; inflorescence bracts not dark and shiny · · · · · · · · · · · 31
31. Lamina smooth, with strongly papillose-dentate margins · · · · · · · · **18.** *dissimilis*
–   Lamina smooth or rugulose, without strongly papillose-dentate margins · · · 32
32. Lamina rugulose throughout· **17.** *subtilis*
–   Lamina smooth, or only basal sheaths rugulose · 33
33. Lateral sepals only dentate towards apex; leaves generally present at flowering · **20.** *capnoides*
–   Lateral sepals with flattened teeth from middle to apex; leaves almost absent at flowering· **19.** *erubescens*
34. Fertile inflorescence bracts more than 10; inflorescence often elongating with age; leaves often spirally twisted · 35
–   Fertile inflorescence bracts fewer than 10; inflorescence not elongating with age; leaves not spirally twisted· 36

35. Outer inflorescence bracts somewhat elongated, thinner and paler than those above, reaching half the length of inflorescence; coastal southern Africa · **8.** *natalensis*
– Outer inflorescence bracts very short, similar in colour and texture to rest, less than ¹/₈ length of inflorescence; widespread · · · · · · · · · · · · · · · · · **6.** *congensis*
36. Inflorescence small, up to 4 mm long; sterile bracts 4; teeth on keel of lateral sepals minute, around middle only; leaves subterete in section · · · **15.** *capillaris*
– Inflorescence at least 5 mm long; keel of lateral sepals toothed for most of length; leaves distinctly flattened · **13.** *elegantula*

1. **Xyris kornasiana** Brylska & Lisowski in Polish Bot. Stud. **1**: 117, figs.1–4 (1990). — Lock in F.T.E.A., Xyridaceae: 4 (1999); in Kew Bull. **54**: 306 (1999). —Lisowski *et al.* in F.A.C., Xyridaceae: 101, fig.43 (2001). Type: Congo, upper Katanga, Kundelungu Plateau, R. Lofoi, 25.v.1984, *Malaisse* 13018 (BR holotype, K).

Perennial herb (but see note below), loosely caespitose. Leaves up to 30 × 0.2 cm; sheaths ± half as long as lamina, brown, strongly rugulose to scabrid, margins thin, smooth, scarious, entire, gradually narrowing upwards; ligule short; lamina linear, flattened, 2-ridged, c.2 mm wide, glabrous, reticulate-rugulose at least between veins; apex acuminate-incurved. Peduncle 35–60 cm long, 1–1.5 mm in diameter, terete, grooved longitudinally, rugulose at least between ridges when dry; sheaths a little shorter than leaves, smooth and brown below, greenish and rugulose above with a short terminal blade. Inflorescence ellipsoid, acute, whitish, to 2 × 1 cm. Sterile bracts membranous, whitish with a pale brownish central region, the outer 2–3 narrowly ovate, acute, up to 20 × 5 mm, the inner to 13 × 5 mm; fertile bracts 18–28, ovate, acute, membranous, whitish. Lateral sepals curved, boat-shaped, acuminate, 8 × 3 mm, keel entire. Corolla yellow to orange, tube 1–1.5 mm long, lobes obovate, 3.5–4 × 1.5–2.2 mm. Stamens 3–3.5 mm long; staminodes 2.8–3 mm long, bifid with tufts of yellow hairs 1.5 mm long at apex. Ovary ellipsoid, 1–2 × 0.5–0.8 mm in diameter, style 2.5–3.5 mm long, trifid above middle. Capsule ellipsoid, c.5 mm long; seeds ellipsoid with 14–15 longitudinal ridges, c.0.5 × 0.25 mm.

Zambia. N: Kasama Dist., Kasama–Mbala (Abercorn) road, c.50 km from Kasama, 8.iv.1961, *Richards* 15036 (K, SRGH). **Malawi**. N: Mzimba Dist., Luwawa Dam, Vipya, 120 km S of Mzuzu, 8.ii.1971, *Pawek* 4401 (K, MAL).
Also in Tanzania and Congo. In roadside ditches, bogs and dambos; 1200–1650 m.
Conservation notes: A fairly widespread taxon but there are few collections in spite of its distinctive appearance, suggesting that it may be genuinely rare; probably Lower Risk near threatened.
The long whitish membranous outer bracts and the rugulose leaves and scapes should readily distinguish this from all other species in the area. The type and *Pawek* 4401 are somewhat less rugulose than *Richards* 15036 and material from Tanzania; there is too little material to decide whether this has any significance.

2. **Xyris aristata** N.E. Br. in F.T.A. **8**: 11 (1901). —Malme in Svensk Bot. Tidskr. **6**: 552 (1912). —Lewis in Fl. Cameroon **22**: 49 (1981). —Lock in F.T.E.A., Xyridaceae: 8 (1999); in Kew Bull. **54**: 306 (1999). —Lisowski *et al.* in F.A.C., Xyridaceae: 5, fig.1 (2001). Type: Zambia, Mbala Dist., Kambole, 1896, *Nutt* s.n. (K holotype).

 Xyris subaristata Malme in Ark. Bot. **22**(4): 5 (1928). —Hepper in Kew Bull. **21**: 425 (1968). —Lisowski *et al.* in F.A.C., Xyridaceae: 7, fig.2 (2001). Type: Congo, Katanga, between Dembo and Welgelegen, 2.v.1912, *Bequaert* 370 (BR holotype).

 Xyris laciniata Hutch., Botanist Sthn. Africa: 490 (1946). —Hepper in Kew Bull. **21**: 425 (1968). Type: Zambia, Kabwe Dist., 50 km NE of Kabwe (Broken Hill), 13.vii.1930, *Hutchinson & Gillett* 3641 (K holotype, BR, LISC).

Xyris dilungensis Brylska in Bull. Jard. Bot. Belg. **60**: 139 (1990). —Lisowski *et al.* in F.A.C.,
Xyridaceae: 11, fig.4 (2001). Type: Congo, Katanga, Kundelungu Plateau, near source of R.
Lofoi, 1600 m, 7.i.1971, *Lisowski* 23000a (POZG holotype, BR).

Perennial herb (but see note below) forming dense clumps including fibrous remains of old
leaf sheaths. Leaves up to 50 × 0.25–0.4 cm; sheaths gradually tapering from base, black to
brown, sometimes splitting into persistent bristles when old, usually rugulose, margins
sometimes ciliate; ligule rounded, 2–3 mm long; lamina linear, minutely papillose, more
strongly so on margins; apex solid, acute-incurved. Peduncle 60–100 cm long, 2–3 mm in
diameter, terete with two longitudinal papillose ridges. Inflorescence subglobose, 0.8–1.5 cm in
diameter, blackish brown. Sterile bracts suborbicular, coriaceous, papillose particularly above,
c.7 × 6 mm including a pale membranous fimbriate margin and a black rigid projecting point;
fertile bracts similar, or more papillose and with a shorter acumen. Lateral sepals c.5 mm long,
curved, papyraceous, dark brown with paler membranous margins and a blackish papillose
acumen, keel irregularly erose-ciliate. Corolla yellow, tube 4 mm long; lobes obovate, 4.5 × 4
mm, apex irregularly toothed. Filament ligulate, 1 mm long, anthers 2 mm long; staminodes
2–3 mm long. Ovary obovoid, 2 × 1 mm; style 3 mm long, 3-branched, each branch 2.5 mm
long, stigmas fimbriate. Capsule not seen; seeds broadly ellipsoid, c.0.5 × 0.35 mm, with c.12
longitudinal ridges joined by weak transverse ridges.

Zambia. N: Mpika Dist., Lake Chibakabaka, 15.x.1963, *Robinson* 5760 (K, SRGH).
W: Mufulira Dist., Mufulira, 8.vi.1934, *Eyles* 8349 (K, SRGH). C: Naluseka R., by Great
North Road, 19 km N of Ndola (Broken Hill), 23.ix.1947, *Brenan & Greenway* 7915
(K). **Malawi.** N: Chitipa Dist., W slopes of Nganda, 23.vii.1972, *Brummitt & Synge*
WC64 (K).

Also in Tanzania and Angola. In grass swamps, streamsides and ditches; 900–2450 m.

Conservation notes: A widespread taxon, although there are some taxonomic
uncertainties; not threatened.

This species is distinctive in its long-aristate bracts. The type is immature and
atypical. Lisowski *et al.* (2001) kept *Xyris subaristata*, *X. laciniata* and *X. dilungensis*
separate, but they are here considered to be extremes within a variable species, which
should also probably include *X. leonensis* Hepper from West Africa (Sierra Leone to
Central African Republic and, according to Lisowski *et al.*, Congo and Burundi).
Walter 21 (K) (Zambia N: Mweru–Luapula, Mbereshi, 9 km from Kazemba's,
iii.1932), with long pale outer bracts, is typical of material that would be placed
under *X. leonensis* by some. *Richards* 163 and 6008 (both N Zambia) represent
intermediate material.

3. **Xyris rehmannii** L.A. Nilsson in Kongl. Svenska Vetenskapsakad. Handl. **24**(14):
   28 (1892). —Malme in Bot. Jahrb. Syst. **48**: 299 (1912); in Bot. Not. **1932**: 11
   (1932). —Hepper in Kew Bull. **21**: 424 (1968); in F.W.T.A. ed.2, **3**(1): 54 (1968).
   —Lewis & Obermeyer in F.S.A. **4**(2): 7 (1985). —Lock in F.T.E.A., Xyridaceae:
   11 (1999); in Kew Bull. **54**: 307 (1999). —Lisowski *et al.* in F.A.C., Xyridaceae: 51,
   fig.22 (2001). Type: South Africa, Gauteng, Houtbosch, 1875–80, *Rehmann* 5764
   (Z holotype, BM, K).
   Xyris rigidescens Rendle in Cat. Afr. Pl. Welw. **2**(2): 67 (1899). —N.E. Brown in F.T.A. **8**: 11
   (1901). Type: Angola, Huíla, around Lopollo, xii.1859, *Welwitsch* 2474 (BM holotype, K).
   Xyris dispar N.E. Br. in F.T.A. **8**: 12 (1901). —Hepper in Kew Bull. **21**: 424 (1968). Type:
   Zimbabwe, near Harare (Salisbury), Six Mile Spruit, xi.1899, *Cecil* 152 (K holotype).

Perennial herb, sometimes forming tussocks. Leaves up to 50 × 0.3 cm; sheaths red-brown,
grooved, minutely rugulose when old and dry; ligule triangular, membranous, 2–3 mm long;
lamina flat, glabrous, smooth; apex solid, acuminate. Peduncle 80–120 cm long, 0.5 mm in
diameter above, 2 mm below, somewhat flattened, particularly above, lightly longitudinally
grooved when dry. Inflorescence subspherical at maturity, 8–12 mm in diameter. Sterile bracts

coriaceous, shiny, brownish black, the outermost sometimes with a distinct dorsal keel, broadly obovate to oblong, 5–7 × 4–5 mm; fertile bracts similar to sterile ones but narrower, numerous, keeled toward apex and apiculate, apiculus often twisted. Lateral sepals curved, boat-shaped, acuminate, keel dark brown, coriaceous and ciliolate-denticulate, excurrent in a projecting point, the margins thinner, paler and translucent. Corolla yellow; tube c.3 mm long, lobes suborbicular, c.3.5 × 4 mm, rounded at apex. Stamens 3 mm long; anthers 2 mm; staminodes conspicuous, bifurcate, the branches with numerous long hairs. Style arms c.3 mm long, stigmas flattened and lobed. Capsule broadly ellipsoid, c.4 mm long. Seeds numerous, ellipsoid, with 16–18 longitudinal ridges, c.0.5 × 0.25 mm.

Zimbabwe. W: Matobo Dist., Farm Besna Kobila, xi.1954, *Miller* 2536 (K, SRGH). C: Goromondzi Dist., Denda Estate, c.30 km E of Harare off Shamva road, 29.xi.1957, *Phipps* 822 (K, LISC, SRGH). E: Nyanga Dist., Nyanga Village, Rochdale R., 10.xi.1978, *Nicholas* 532 (K, SRGH). S: Masvingo Dist., 8 km E of Morgenster Mission, S of Lake Mutirikwi (Kyle), 18.xii.1970, *Müller & Pope* 1717 (K, SRGH). **Malawi**. N: Nkhata Bay Dist., 110 km S of Mzuzu, Viphya, Luwawa Dam, 8.ii.1971, *Pawek* 4406 (K, MAL).

Also known from Kenya, Tanzania, Angola, Congo and South Africa. Riverbanks, streamsides, permanent swamps and other wet places in higher rainfall areas; 1300–1800 m.

Conservation notes: Widespread and frequent in its habitat; not threatened.

A robust perennial with almost black shiny inflorescence bracts.

4. **Xyris gerrardii** N.E. Br. in Fl. Cap. **7**: 5 (1897). —Malme in Bot. Jahrb. Syst. **48**: 299 (1912); in Bot. Not. **1932**: 12 (1932). —Lewis & Obermeyer in F.S.A. 4(2): 7 (1985). —Lock in F.T.E.A., Xyridaceae: 10 (1999); in Kew Bull. **54**: 308 (1999). Type: South Africa, KwaZulu-Natal, 'Zululand', 1865, *Gerrard* 1526 (K holotype, BM, NH).

Perennial herb forming clumps of short stout horizontal or ascending rhizomes covered in leaf sheaths. Leaves up to 25 cm × 0.5–1 mm; sheaths much shorter than blades, gradually dilated towards base, dark brown with paler margins, smooth, sometimes verrucose below, tapering to a small ligule; lamina linear, flattened, glabrous, longitudinally furrowed, tending to be spirally twisted; apex acute. Peduncle up to 37 cm long, 0.2–0.5 mm in diameter, 2-ridged, particularly above, glabrous. Inflorescence ellipsoid to obconical, c.7 mm long, 4–5 mm in diameter. Sterile bracts broadly ovate, acute, c.3 × 2.5 mm, coriaceous, blackish-brown with a broad scarious paler margin, becoming lacerate with age. Fertile bracts similar but narrower, keeled towards apex, mucronate. Lateral sepals c.4.5 mm long, keeled, margins membranous, laciniate, keel finely and bluntly denticulate, often only below, excurrent above in a mucro. Corolla yellow; tube c.4 mm long, lobes broadly obovate, c.3 × 2 mm, entire; staminodes bifurcate, each branch with a bunch of long hairs. Stamens with filaments c.0.5 mm long; anthers 2 mm long. Capsule oblong-ovoid. Seeds not seen.

Zimbabwe. E: Nyanga Dist., near Mt Nyangani summit ridge, 12.xii.1963, *Whellan* 2073 (SRGH). **Mozambique**. MS: Mt Gorongosa, Gogogo summit area, 13.xii.1963, *Tinley* 2223 (SRGH).

Also in Tanzania and South Africa. Moist or boggy places in upland grasslands; 2000–2500 m.

Conservation notes: An uncommon species restricted to moist places in upland grasslands; easily overlooked. The small number of collections and the restricted habitat suggest Lower Risk near threatened.

A small clump-forming perennial with dark inflorescence bracts with scarious margins, occurring in upland grasslands. It resembles *X. obscura* but differs in the scarious margins of the inflorescence bracts.

5. **Xyris obscura** N.E. Br. in F.T.A. **8**: 16 (June 1901). —Hepper in Kew Bull. **21**: 424 (1968). —Lock in F.T.E.A., Xyridaceae: 10 (1999); in Kew Bull. **54**: 308 (1999). —Lisowski *et al.* in F.A.C., Xyridaceae: 80, fig.33 (2001). Type: Zimbabwe, near Harare (Salisbury), Six Mile Spruit, xi.1899, *Cecil* 152A (K holotype).

> *Xyris brunnea* L.A. Nilsson in Bot. Jahrb. Syst. **30**: 271 (July 1901). —Malme in Bot. Jahrb. Syst. **48**: 300 (1912); in Bot. Not. **1932**: 12 (1932). Type: Tanzania, Livingstone Mts, W of Ubena, iii.1899, *Goetze* 822 (B holotype, BM, EA, K).
>
> *Xyris nivea* sensu Lewis in Fl. Cameroon **22**: 40 (1981) and Lewis & Obermeyer in F.S.A. **4**(2): 5 (1985), non Rendle (see note in F.T.E.A.).

Tufted perennial herb forming dense clumps, often bulked out by persistent leaf bases; shoot bases sometimes somewhat swollen and bulbous. Leaves to 15 × 0.1 cm; sheaths to 0.8 cm wide at base, smooth and with ciliate margins below, becoming rugulose above when dry, dark red-brown in centre, paler at margins; ligule not seen; lamina terete to semiterete, glabrous, rugulose; apex aristate. Peduncle 13–28 cm long, c.0.7 mm in diameter, terete, longitudinally ridged. Inflorescence 3–5-flowered, ellipsoid, almost black, 5–6 mm long and 2.5 mm in diameter. Sterile bracts broadly obovate, apiculate, 4 × 3 mm, minutely rugulose on back, black with very narrow paler border. Fertile bracts broadly obovate, strongly concave, apiculate, 5 × 4 mm, minutely rugulose on back, black above, paler below. Lateral sepals curved, c.5 mm long, laciniate on keel above; third sepal c.5 mm long. Corolla tube c.1.5 mm long; lobes obovate, c.4 × 3 mm, apex emarginate, irregularly laciniate. Filament ligulate, c.0.5 mm long, anthers 1.5 mm long; staminodes much-branched, c.1 mm long. Style c.3 mm long, branches 2 mm long; stigmas much-branched. Seeds broadly ellipsoid, 0.6 × 0.35 mm, with more than 25 interconnected longitudinal ridges.

Zambia. N: Isoka Dist., Mafinga Hills, 12.iii.1961, *Robinson* 4464 (K). **Zimbabwe**. E: Mutare Dist., Vumba Mts, Summit Castle Beacon, 18.iii.1956, *Chase* 6027 (K, SRGH). **Malawi**. N: Chitipa Dist., Mafinga Mts, near highest point at N end of ridge, 2.iii.1982, *Brummitt, Polhill & Banda* 16264a (K, LISC). S: Zomba Dist., Zomba Plateau, below road to summit opposite Malosa Saddle, 15.iii.1970, *Brummitt* 9144 (K, SRGH). **Mozambique**. MS: Sussundenga Dist., Tsetserra, 7.ii.1955, *Exell, Mendonça & Wild* 249 (LISC, SRGH).

Also known from Tanzania, Congo, Angola and South Africa. Usually in moist places in upland grasslands; 1600–2600 m.

Conservation notes: Occurs in similar habitats to *Xyris gerrardii* but is much more frequently collected; not threatened.

The very convex black inflorescence bracts, which are usually aristate and without scarious margins, are distinctive.

The type is labelled as being collected at 'Six Mile Spruit', near Harare. No other material has been seen from this area or altitude, so there must be some doubt as to the correctness of the label data.

Lewis regarded *Xyris nivea* Rendle as the correct name for this taxon. The type of *X. nivea* (*Welwitsch* 2468 (BM holotype)) is, however, very different; it lacks a bulbous base and the inflorescence bracts are thinner, paler, and lack the papillae and apiculus of *X. obscura*. The plants from Cameroon identified as *X. nivea* by Lewis (1981) also appear to belong to a different and probably unnamed taxon.

Lisowski (2001) records this species from Burundi. I have seen much of the cited material, which in my opinion belongs to a different taxon.

6. **Xyris congensis** Büttner in Verh. Bot. Vereins Prov. Brandenburg **31**: 71 (1890). —N.E. Brown in F.T.A. **8**: 23 (1901). —Malme in Bot. Jahrb. Syst. **48**: 298 (1912). —Lewis & Obermeyer in F.S.A. **4**(2): 4 (1985). —Lock in F.T.E.A., Xyridaceae: 8, fig.3 (1999); in Kew Bull. **54**: 309 (1999). —Lisowski *et al.* in F.A.C., Xyridaceae: 29, fig.12 (2001). Type: Congo, between Lukolela and Equator Station, 10.xi.1885, *Büttner* 583 (B holotype). FIGURE 13.4.**1**.

Xyris hildebrandtii L.A. Nilsson in Öfvers. Förh. Kongl. Svenska Vetensk.-Akad. **1891**(3): 155 (1891). —N.E. Brown in F.T.A. **8**: 24 (1901). —Malme in Bot. Jahrb. Syst. **48**: 295 (1912); in Bot. Not. **1932**: 11 (1932). —Hepper in Kew Bull. **21**: 421 (1968). Type: Madagascar, East Imerina, Andrangoloaka, xi.1880, *Hildebrandt* 3724 (M holotype, K).

Xyris hildebrandtii var. *angustifolia* Malme in Svensk Bot. Tidskr. **24**: 393 (1928). Type: "cotypus" [in Malme's hand], Malawi, Mt Mulanje, *Adamson* 414 (K).

Xyris nitida L.A. Nilsson in Öfvers. Förh. Kongl. Svenska Vetensk.-Akad. **1891**(3): 156 (1891). —N.E. Brown in F.T.A. **8**: 24 (1901). —Malme in Bot. Jahrb. Syst. **48**: 297 (1912). Type: Equatorial Guinea, Corisco Is., 1862, *Mann* 1858 (B holotype, K, P).

Xyris umbilonis L.A. Nilsson in Kongl. Svenska Vetenskapsakad. Handl. **24**(14): 30 (1892). —Rendle in Cat. Afr. Pl. Welw. **2**(1): 67 (1899). —Malme in Bot. Jahrb. Syst. **48**: 296 (1912). Type: South Africa, KwaZulu-Natal, Umbilo, 1875–80, *Rehmann* 8139 (Z holotype, K).

Xyris batokana N.E. Br. in F.T.A. **8**: 23 (1901). —Malme in Bot. Jahrb. Syst. **48**: 296 (1912). Type: Zambia, Batoka Highlands, vii.1860, *Kirk* s.n. (K holotype).

Xyris baumii L.A. Nilsson in Warburg, Kunene-Sambesi Exped. Baum: 181 (1903). —Malme in Bot. Jahrb. **48**: 297 (1912). Type: Angola, Kuebe, above mouth of Rio Kubango, 1150 m, 28.x.1899, *Baum* 333 (BR, K isotypes).

Xyris leptophylla Malme in Svensk Bot. Tidskr. **6**: 549 (1912). —Lisowski *et al.* in F.A.C., Xyridaceae: 93, fig.40 (2001). Type: Congo, Katanga, Bulelo R., ix.1911, *Fries* 517 (UPS holotype).

Xyris leptophylla var. *subbatokana* Malme. This name was written by Malme on several sheets from Zambia at K (*Allen* 483, *Fries* 853a, 853g, *Rogers* 8375), but I can find no evidence that it was ever published.

Xyris extensa Malme in Ark. Bot. **22**(4): 4 (1928). Types: Congo, Wombali, vi.1913, *Vanderyst* 1172; Wombali, viii.1913, *Vanderyst* 2019 [not 2119]; Apundu, 1913, *Bavicchi* 340; Atené, i.1914, *Vanderyst* 3128; Luebo-Kasai Region, ix.1915, *Achten* 236B (all BR syntypes).

Perennial herb, often with base bulked out by persistent leaf bases, forming clumps. Leaves up to 30 × 0.1–0.6 cm; sheaths red-brown, shiny, glabrous, gradually dilated to base, splitting and persisting as bristles when old; ligule acute, c.1 mm long; lamina linear, glabrous, often spirally twisted; apex solid, acute, glabrous. Inflorescence ellipsoid, obtuse, c.9 × 6 mm in diameter at first flowering, elongating with age to c.20 × 6 mm or longer. Sterile bracts suborbicular, 3–4 × 3–4 mm, coriaceous, shiny, uniformly brown, usually without a dorsal mark; fertile bracts similar, numerous. Lateral sepals curved, 3.5–4.5 mm long, coriaceous, brown, keel irregularly fimbriate-dentate in upper half. Corolla yellow; petals broadly obovate, apex lacinate, 4.5 × 3.5 mm. Stamens 2 mm long including anthers to 1 mm; staminodes almost sessile, composed of bunches of straight hairs. Ovary ellipsoid, style c.2.5 mm long, branches 2 mm long. Seeds ovoid, c.0.6 × 0.25 mm, with 12–13 longitudinal ridges.

Zambia. B: Kalabo Dist., 5 km W of Kalabo, 16.xi.1959, *Drummond & Cookson* 6533 (K, LISC, SRGH). N: Mansa Dist., Chipili, 60 km N of Mansa (Fort Rosebery), 27.vi.1956, *Robinson* 1757 (K, SRGH). W: Ndola Dist., Luanshya, 28.ii.1960, *Robinson* 3360 (K, SRGH). C: Mpika Dist., Mutinondo Wilderness Area, 14.vi.1998, *P.P. Smith* 1705 (K). E: Petauke Dist., Kacholola, 21.x.1967, *Mutimushi* 2182 (K, NDO). S: Choma Dist., near Choma, Siamambo Forest Reserve, by Siamambo stream, 23.vii.1952, *Angus* 15 (FHO, K). **Zimbabwe**. N: Gokwe Dist., Mafungabusi Plateau by source of Sikombella stream, 25.vi.1947, *Keay* in *FHI* 21368 (FHO, K). W: Matobo Dist., Farm Besna Kobila, v.1957, *Miller* 4376 (K, LISC, SRGH). C: Harare (Salisbury), 26.xii.1931, *Eyles* 7069 (K, SRGH). E: Nyanga Dist., Pamushana Farm, xi.1957, *Miller* 4663 (K, SRGH). **Malawi**. N: Mzimba Dist., 5 km W of Mzuzu, Katoto, 3.viii.1973, *Pawek* 7280 (EA, K, MAL, MO, SRGH). C: Mchinji Dist., 5 km N of Mchinji on road to Kasungu, 27.iv.1970, *Brummitt* 10217 (K, LISC, SRGH). S: Zomba Dist., Zomba Plateau, Mlunguzi Marsh, 9.i.1992, *Goyder, Paton & Kaunda* 3503 (BR, K, LMU, MAL, MO, PRE). **Mozambique**. N: Muecate Dist., Imala, on road to Mecuburi, 21.xi.1936, *Torre* 1068 (LISC). Z:

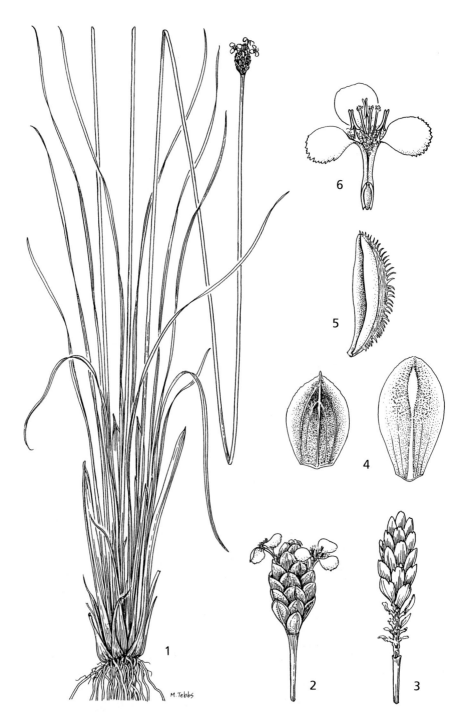

Fig. 13.4.**1**. XYRIS CONGENSIS. 1, habit (× ²/₃); 2, young spike (× 1); 3, old spike (× 1); 4, bracts (× 8); 5, lateral sepal (× 8); 6, flower (× 4). All from *Verdcourt* 2791. Drawn by Margaret Tebbs. From Flora of Tropical East Africa.

Gurué Dist., Mt Namuli, Muretha Plateau, 26.v.2007, *Harris et al.* 184 (K). MS: Mossurize Dist., Makurupini R. just below Makurupini Falls, 22.iv.1973, *Ngoni* 208 (K, LISC, SRGH). M: Matutuíne Dist., between Zitundo and Ponta do Ouro, 3.xii.1968, *Balsinhas* 1416 (LISC).

Widespread in tropical Africa from Uganda west to Gabon and Cameroon and south to South Africa, also in Madagascar. Wet grasslands and seasonal and perennial swamps; 300–2300 m.

Conservation notes: Widespread and often frequent where it occurs; not threatened.

The spiralling of the leaves and the inflorescences which elongate with age and from which the lower bracts and flowers fall, together with the rounded, brown and usually glossy bracts, usually allow easy recognition of this perennial clump-forming species.

As treated here this is a variable taxon, as can be seen from the synonymy; Lewis (1981) came to similar conclusions. Variation occurs in several characters but these do not appear to be correlated. Leaf width varies considerably and there has been a tendency to call the broader-leafed forms *Xyris hildebrandtii* and the narrowest *X. leptophylla*. Material of *X. extensa* has even broader leaves and often shows elongated rhizomes. The type of *X. extensa* also has an elongated inflorescence (see *X. baumii* below). The fertile bracts may be papillose to varying degrees; the extreme is represented by *Greenway & Trapnell* 5550 (EA, K, LISC, SRGH) in which the bracts are densely and heavily papillose and also rather dark and opaque. Plants with papillose bracts are found throughout the area with a concentration in northern Zambia. Plants with a dorsal mark (developed to varying degrees) on the fertile bracts tend to be from Zimbabwe, but are also found in Zambia and Angola; there are usually collections with unmarked bracts from the same or nearby localities. Elongation of the lowest sterile bracts and development of an acumen tends to be most pronounced in material from higher altitudes in Malawi, but similar specimens are found elsewhere. Specimens with more than 4 sterile bracts are found throughout the region. All these variants also occur in Angola. They may occur in combination, for example the type of *X. batokana* has more than 4 sterile bracts and weakly papillose fertile bracts with a slight dorsal mark. The type of *X. baumii* is rather similar but the inflorescence is much more elongated and the lateral sepals project beyond the fertile bracts. Again these are characters which vary over the range of the taxon without clear or consistent correlation with others. Such variation makes the recognition of varieties, at first sight a tempting option, of doubtful value.

Milne-Redhead 4439 (Zambia W: Mwinilunga, Kabompo R. (K, LISC)) is clearly part of the same complex but has elongated horizontal and vertical rhizomes. This may be a result of seasonal submergence in the riverside habitat noted by the collector. *Richards* 17423 (K) from the same area and habitat may be the same but lacks the basal parts. Both are relatively robust plants and approach the type of *X. extensa*.

Lisowski (2001) rejects my earlier placement of *X. leptophylla* in the synonymy of *X. congensis*. I have examined some of the material that he cites (e.g. *Lewalle* 4953 and *Lewalle* 4873), and I remain of the opinion that these specimens represent only the narrow-leaved end of a spectrum of variation. The other character that he mentions, the degree of ridging on the peduncle, is also variable and may partly be a function of the stage of maturity and of the speed of drying.

X. gossweileri, *X. foliolata* and *X. natalensis* appear to represent more extreme and consistent variants of this same complex, but are treated here as good species.

7. **Xyris gossweileri** Malme in Ark. Bot. **24**(5): 3 (1932). —Lock in Kew Bull. **54**: 310 (1999). —Lisowski *et al.* in F.A.C., Xyridaceae: 27, fig.11 (2001). Type: Angola, Menongue, Caranda, 5.i.1906, *Gossweiler* 2471 (BM, K, LISC isotypes).

Perennial caespitose herb; base of stems somewhat bulbous, covered by persistent leaf sheaths. Leaves not always present at anthesis, 30–40 × 0.1–0.15 cm; sheaths 7–9 cm long, glabrous, broader than lamina above, expanded below, dark chestnut to almost black, shiny; ligule distinct; lamina terete at base, flattened above, linear, smooth but puncticulate when dry, glabrous; apex acute, solid, slightly flattened, sometimes (?always) with a caducous apical projecting point. Peduncle upright, to 70 cm long and c.1 mm diameter, smooth but puncticulate when dry; basal sheaths c.12 cm long. Inflorescence ellipsoid, 10–12 × c.5 mm. Sterile bracts broadly ovate, c.3 mm long, coriaceous, entire, chestnut, shiny, apex truncate-rounded, without keels; intermediate (fertile) bracts broadly ovate to obovate, up to 6 × 4 mm, convex, without keels, rounded at apex; uppermost (fertile) bracts weakly keeled towards apex. Lateral sepals falcately curved, almost linear, 4.5–5 × c.0.8 mm, apex obtuse, brownish; keel narrow, with prickle-like hairs in upper $^1/_2$ to $^2/_3$, not excurrent. Seeds broadly ellipsoid, 0.3–0.4 × 0.2–0.25 mm, with numerous indistinct longitudinal ridges.

Zambia. B: Mongu Dist., Loma Pan, 2.iv.1964, *Verboom* 1701 (K, SRGH). N: Kasama Dist., 100 km E of Kasama, 4.iii.1961, *Robinson* 4430 (K). W: Mwinilunga, 15 km W of Kalene Hill, 14.xii.1963, *Robinson* 6035 (BR, K, SRGH).

Also in Angola and Congo. Seasonally and permanently wet grasslands and dambos, often burned during the dry season; 1100–1400 m.

Conservation notes: Not commonly collected, perhaps because of its relatively precocious flowering; probably Vulnerable.

Inflorescences in this species are very similar to those of *X. congensis*, although usually larger, but the broad black shiny leaf sheaths and more bulbous base are distinctive.

Angus 2745 from Zambia, W of Kawambwa, 7.iv.1961 (K, LISC), differs in the pale-margined bracts which tend to be arranged in vertical rows within the inflorescence, and in the minutely rugulose peduncles. Material at EA under *Angus* ZM 63 with the same date and locality appears to be the same collection. More material is needed.

Xyris bampsii Lisowski is superficially similar but has flattened leaves that are minutely fimbriate-dentate at the margins, particularly towards the apex. It is known from SE Congo and should be looked for in NW Zambia.

8. **Xyris natalensis** L.A. Nilsson in Öfvers. Förh. Kongl. Svenska Vetensk.-Akad. **1891**(3): 157 (1891). —N.E. Brown in Fl. Cap. **7**: 4 (1897). —Malme in Bot. Jahrb. Syst. **48**: 296 (1912). —Lewis & Obermeyer in F.S.A. **4**(2): 4 (1985), excluding *X. foliolata*, here treated as a good species. —Lock in Kew Bull. **54**: 311 (1999). Types: South Africa, KwaZulu-Natal, Natal Bay, 1860, *Sanderson* 455 (K, S syntypes); Cape of Good Hope, *Wahlberg* s.n. (S syntype); Durban (Port Natal), flats between Umlazi R. and Durban, xi.1839, *Krauss* 141 (BM, K, M, Z syntypes).

Tufted perennial herb. Rhizome horizontal or ascending, covered by persistent shiny brown leaf bases and sheaths. Leaves up to 50 × c.0.1 cm in diameter, glabrous; sheaths 5–8 cm long, blackish below, red-brown above, in 2 rows, shiny, appressed red-brown hairy at extreme base, otherwise glabrous, smooth; ligule rounded, c.1 mm long; lamina terete, to 50 × 0.1 cm, glabrous, apex solid, acute. Peduncle 50–80 cm long, 1–2 mm in diameter, thin, wiry, 3-angled below, terete above with 3–4 longitudinal ridges, basal sheath 6–12 cm long, chestnut-brown. Inflorescence broadly ovate or oblong, 10–28 × 7–8 mm. Sterile bracts narrowly ovate, 5–6 × 3 mm, usually spreading, smooth, sometimes aristate; fertile bracts broadly elliptic, 4 × 3 mm, obtuse, punctate above but not forming a distinct dorsal mark. Lateral sepals oblong, curved, c.5 mm long, keeled, the keel broadly winged, the wing margin hispid-spinulose. Corolla yellow. Capsule cylindrical, 3 mm long. Seeds (from South African material) ellipsoid, c.0.5 × 0.2 mm, with 11–14 longitudinal ridges joined by inconspicuous cross bars.

Mozambique. M: Matutuíne Dist., Ponta do Ouro, 4 km along road to Maputo, 3.i.1980, *de Koning* 7890 (BR, K).

Also in South Africa (coastal KwaZulu-Natal). Moist coastal grasslands and swamps; near sea-level.

Conservation notes: Moist coastal grasslands are a habitat vulnerable to cultivation and development for tourism; possibly Vulnerable.

Very similar to *Xyris congensis,* of which it may be no more than a local variant. The slightly elongated outer bracts are usually distinctive, although this feature is shared with the next species.

Lewis & Obermeyer (1985) treated the Munich duplicate of *Krauss* s.n. as the 'putative lectotype'. Although Nilsson states that he saw this duplicate "…in Herb. Monach.", this does not necessarily justify its selection over the other cited material.

9. **Xyris foliolata** L.A. Nilsson in Kongl. Svenska Vetenskapsakad. Handl. **24**(14): 65 (1892). —N.E. Brown in F.T.A. **8**: 10 (1901). —Malme in Bot. Jahrb. Syst. **48**: 295 (1912). —Lock in Kew Bull. **54**: 311 (1999). —Lisowski *et al.* in F.A.C., Xyridaceae: 23, fig.9 (2001). Type: Angola, Malange, x.1879, *Mechow* 277 (Z holotype).

Caespitose perennial herb. Leaves up to 40 cm long; sheaths red-brown, darker and hairy towards base, white-puncticulate; lamina subterete, to 40 × 0.1–0.15 cm, glabrous, ridged longitudinally when dry; apex acute, solid. Peduncle c.60 cm long, terete, longitudinally ridged, glabrous, with basal glabrous sheath 3.5 cm long. Inflorescence ovoid, 1.2–1.4 × 0.7–0.8 cm. Sterile basal bracts narrowly ovate-triangular, spreading, 8–10 × 2–4 mm; fertile bracts broadly obovate, 5–7 × 3.5 mm, coriaceous, concave, entire; apex obtuse, shortly apiculate, punctate on dorsal surface forming a subterminal triangular mark. Lateral sepals oblong to narrowly obovate, 2.5–4 × 0.5–0.6 mm, keeled, keel narrowly winged, with spreading hairs ± along the whole length, excurrent as an apiculus. Corolla yellow, 3.5–4.5 mm long, lobes obovate, 2.5–3 mm long, dentate at apex. Stamens 2.5–3.5 mm long; anthers linear, twice as long as broad; filaments very short; staminodes 1.6–2.8 mm long, bifurcate, branches ending in tuft of yellow hairs. Ovary 1–1.8 mm long; style 1.2–2.4 mm long, divided into 3 above middle, branches apically dilated. Seeds narrowly obovoid or fusiform, with several spiral longitudinal ridges.

Zambia. B: Senanga Dist., 16 km N of Senanga, 31.vii.1952, *Codd* 7280 (SRGH). W: Mwinilunga Dist., Lisombo R., 8.vi.1963, *Loveridge* 856 (BR, K, P, SRGH).

Also in Congo and Angola. Seasonally or permanently wet grassland; c.1400 m.

Conservation notes: Neither common nor widespread; the cited specimens are the only ones seen from the Flora area. Data Deficient, but possibly Vulnerable.

Similar to *Xyris natalensis,* into which it was sunk by Lewis & Obermeyer (1983), and part of the *X. congensis* complex, although the outer bracts are consistently much longer which distinguishes it from *X. congensis.* It seems unlikely that the same taxon should occur both in the coastal wetlands of the Tongaland–Pondoland Regional Mosaic (sensu White 1983) and on the Zambezi–Congo watershed; I prefer to keep them separate on the basis of the key characters.

Details of flowers and seeds taken from Flore de l'Afrique Centrale (Lisowski *et al.* 2001).

10. **Xyris friesii** Malme in Svensk Bot. Tidskr. **6**: 556 (1912). —Lock in Kew Bull. **53**: 892 (1998); in Kew Bull. **54**: 311 (1999). —Lisowski *et al.* in F.A.C., Xyridaceae: 57, fig.25 (2001). Type: Zambia, between Kabwe (Broken Hill) and Chirukutu, 10.viii.1911, *Fries* 296 (UPS holotype, B, BM).

Xyris pocockiae Malme in Ark. Bot. **24**(5): 7 (1931) as *pocockii.* Type: Zambia, near Angola border, 11.v.1925, *Pocock* 252 (B holotype).

Perennial caespitose herb. Leaves 12–20(25) × (0.1)0.15–0.2 cm; sheaths dull green above, sometimes purplish, transversely rugulose, expanded below, brown and opaque, often with long dark brown hairs; ligule absent; lamina narrowed above, transversely rugulose, weakly striate along nerves, apex obtuse. Peduncle subterete, 30–40 cm tall, 0.6–1 mm in diameter, smooth and glabrous, weakly striate, brown to chestnut below; basal sheath 6–10(12) cm long, dark green and minutely tuberculate or transversely rugulose above, brownish or pale chestnut and smooth below. Inflorescence obovoid, 5–8 × 3–5(6) mm, becoming subspherical at maturity. Lowest (sterile) bracts ovate, 3.5–4.5 × 2.5–3 mm, subcoriaceous, whitish or pale yellowish, weakly keeled below rounded apex; fertile bracts obovate to ovate, 5–7 × 3.5–4 mm, somewhat concave, entire, membranous, white-subhyaline to pale yellowish, weakly keeled below rounded often pink-purple apex, without a dorsal mark. Lateral sepals free, almost straight, lanceolate, 5–7 × 1 mm, acute, subhyaline, pale yellow, often tinged pink-purple at apex, keel narrow, shortly and sparsely scabrid-ciliate in lower two-thirds. Corolla yellow; lamina obovate to suborbicular, 4–5 mm long, serrate-lacerate at apex, broadly cuneate at base, claw linear, c.5 × 0.5 mm. Anthers ± linear, c.1.5 mm long; filaments sublinear; staminodes bifurcate, branches with numerous usually moniliform hairs. Ovary unilocular with parietal placentation; ovules numerous; style trifid above middle; styles flabellate-lacerate. Capsule and seeds not seen.

Zambia. N: Mpika Dist., Lukulu R., 16.vii.1930, *Hutchinson & Gillett* 3740 (K). W: Ndola Dist., 16 km S of Ndola, 2.iv.1960, *Robinson* 3452 (BR, K, SRGH). C: Ndola Dist., between Ndola (Broken Hill) & Chirukutu, 10.viii.1911, *R.E. Fries* 296 (B, UPS).

Also in Angola and Congo; c.1300m.

Conservation notes: Infrequently collected and perhaps genuinely rare; probably Lower Risk near threatened.

Distinct in its pale smooth outer inflorescence bracts, lateral sepals with very slightly toothed keels, and rugulose leaves. Plants with young inflorescences are distinctive but older inflorescences are broader and the bracts sometimes darker. The pink-purple tips to the young bracts and sepals seem to be a consistent feature.

Robinson 5703 (EA, K, SRGH) from C Zambia is similar vegetatively but has much darker bracts and a few small spinules near the middle of the keel of the lateral sepals. It may be a variant or a new taxon; more material is needed. *Hooper & Townsend* 339 (K) from W Zambia, Mwinilunga, Kalene Hill, is also rather similar but has a more bulbous base.

Xyris pocockiae almost certainly belongs here. The leaf surface is similarly rugulose, and the inflorescences are very similar, although the outer bracts seem a little more delicate in texture. When Malme described the species he must have been unaware that the collector, Mary Agard Pocock (1886–1977), was female.

Lisowski (2001) records this species from Burundi. In my opinion the specimen cited (*Michel* 1921) does not belong to this species. I have seen specimens from Congo that undoubtedly belong here, but some of the material cited by Lisowski belongs to other taxa.

The type sheet of *Xyris welwitschii* Rendle (*Welwitsch* 2465 (BM)) includes a single plant of *X. friesii*. Some authors, particularly Lewis in Fl. Cameroun, have applied the name *X. welwitschii* to the present taxon. See Lock (1998) for a full discussion and leptotypification.

11. **Xyris lejolyana** Lisowski in Syst. Geogr. Pl. **69**: 209, fig.3 (1999), as 'lejolyanus'. — Lisowski *et al.* in F.A.C., Xyridaceae: 59, fig.26 (2001). Type: Congo, Katanga, 67 km E of Mitwaba, *Duvigneaud & Timperman* 2722 (BRLU, see note below).

Perennial caespitose herb to 90 cm tall, with old leaf sheaths at base. Outer leaves linear, flat, 20–50 × 0.2–0.5 cm, acute at apex; leaf sheath 13–20 × 1–1.4 cm, margin hyaline, not ciliate, weakly papillose; ligule 2 mm long; inner leaves each subtending a flowering culm, 2.5–3.4 cm long; lamina linear, flat, 0.5–3.5 × 0.1–0.18 cm, acute at apex, rugulose; margin papillose; sheath

24.5–30.5 × 0.4–0.5 cm, margined and keeled; ligule short. Peduncle subcylindric, 2-ridged, 60–90 cm long and 0.1–0.3 cm in diameter, glabrous or papillose on ridges. Young inflorescence ellipsoid, later spherical, to 8–14 mm long and wide, 16–31-flowered. Sterile bracts 4, deep yellow, oval to orbicular, 5–7 × 4–6mm, smooth, glabrous, 3–5-nerved, apex rounded; fertile bracts 19–31, deep yellow, almost orbicular to broadly oblong, 6–7 × 5–6 mm, entire, smooth, glabrous, indistinctly 5–7-nerved, apex rounded. Lateral sepals asymmetric, oblanceolate, 6–7 × 1.5–2 mm, apex obtuse; dorsal keel winged, often with 1–3 irregular teeth towards apex. Dorsal sepal narrowly elliptic, 6–7 × 2 mm, yellow, reddish towards the obtuse apex. Corolla yellow, tube 2 mm long, lobes 3.5–4 mm long, oblong, apex irregularly dentate. Stamens 4.5–5 mm long; anthers yellow; filaments brown; staminodes shorter than stamens, bifid, ending in tufts of yellow hairs. Gynoecium 5–5.5 mm long; ovary ellipsoid; style 3–3.5 mm long, trifid above middle. Capsule obovoid to ellipsoid, 3–5 mm long, apiculate at apex, 3-lobed, smooth, glabrous, brown. Seeds ellipsoid, longitudinally ridged, brown, blackish-brown at the apiculate apex.

Zambia. N: Mporokoso Dist., 45 km E of Mporokoso, 13.iv.1961, *Phipps & Vesey-Fitzgerald* 3130 (BR, P). There is an un-numbered collection at K which seems almost certain to be a duplicate.

Also in Congo. In *Protea* zone of dambos; c.1400m.

Conservation notes: Data Deficient, but possibly Vulnerable.

The Zambian material is extremely similar to specimens cited by Lisowski (2001) although the ridges on the peduncle are only slightly papillose, not markedly so as in the Congo material.

The number *Duvigneaud & Timperman* 2722 also appears on sheets of *Xyris aristata* and *X. obscura* because it is a collecting site (relevé) number, not a unique specimen number. I am grateful to P. Bamps for this information.

12. **Xyris symoensii** Brylska & Lisowski in Syst. Geogr. Pl. **69**: 212, fig.5 (1999). — Lisowski *et al.* in F.A.C., Xyridaceae: 61, fig.27 (2001). Type: Congo, Katanga, Kundelungu Plateau, 3 km N of W source of Lutishipuka R., *Lisowski* 90593 (POZG holotype).

Perennial caespitose herb to 75 cm high; base bulbous, surrounded by old brown leaf sheaths; root system in bundles. Leaves in a basal clump, the outer ones subcylindric to weakly flattened, 20–28 × 0.1–0.17 cm, smooth, glabrous; sheaths 7–10 cm long; margins hyaline; ligule absent. Inner leaves 15–20 cm long; sheaths 13–17 × 0.2–0.3 cm, glabrous, smooth, without a ligule; margins hyaline; lamina subcylindric, 2–3 cm long, glabrous, apex acute. Peduncle cylindrical or nearly so, 43–75 cm long, 0.1–0.15 cm in diameter, smooth. Inflorescence ellipsoid, 7–8 × 5–6 mm, the oldest sub-spherical, to 8–10 × 8–10 mm; bracts deep yellow. Sterile bracts 4, ± decussate, elliptic or oblong-elliptic, 5.5–6 × 3.5–4.5 mm, smooth, obscurely 3–5-nerved, apex rounded; margins entire; fertile bracts 9–13, elliptic to obovate, 6.5–7 × 4–5 mm, obscurely 7–9-nerved, apex rounded, margins entire. Lateral sepals 2, linear, 6.5–7 × 1–1.5 mm, asymmetrical; dorsal keel narrowly winged, the wing finely ciliate in middle, apex obtuse; dorsal sepal elliptic, 6–6.5 × 1.5–2 mm, glabrous, reddish, apex acute. Corolla yellow, tube 2 mm long, lobes 3.2–3.5 × 2.5–2.8 mm, obovate; apex irregularly toothed. Stamens 3.5–4 mm long; filaments 1.5–2 mm long, brown; anthers 2 mm long, oblong, yellow; staminodes shorter than stamens, 3–3.5 mm long, bifid, branches ending in tuft of yellow hairs 1.5–1.8 mm long. Gynoecium 4.5–5 mm long; ovary ellipsoid, 1.5 × 0.6 mm, glabrous, smooth, brown; style 3–3.5 mm long, divided above middle into 3 flattened branches. Fruits obovoid to oblong, 3–4 × 2–2.5 mm, 3-lobed, smooth, glabrous, apiculate, brown. Seeds ellipsoid to obovoid, apiculate, longitudinally furrowed, brown, black at apex.

Zambia. N: Kawambwa Dist., Kawambwa–Mbereshi road, 18.iv.1957, *Richards* 9336 (K). W: Mwinilunga, on Kalahari sand, 16.v.1969, *Mutimushi* 3334 (K, LISC, NDO).

Only known from Congo and N Zambia. In moist valleys; 1260–1400 m.

Conservation notes: Few collections from the Flora area; probably Vulnerable.

Xyris symoensii is usually a fairly vigorous species with spherical golden-yellow inflorescences. It is very similar to *X. lejolyana* but the leaves and peduncles are smooth, not rugulose.

13. **Xyris elegantula** Malme in Ark. Bot. **24**(5): 5 (1932). —Lock in Kew Bull. **54**: 312 (1999). —Lisowski *et al.* in F.A.C., Xyridaceae: 45, fig.19 (2001). Type: Angola, Muene Chipipa, rio Kuiriri, iii.1906, *Gossweiler* 2843 (B, BM, K, LISC).

Tufted perennial from a short upright rhizome. Leaf sheaths 2.5–3 cm long, reddish brown, broadened towards base, somewhat transversely rugulose, margins hairy in the type but not in the Flora area; ligule distinct; lamina linear, 5–7 × 0.1–0.15 cm, narrowed to a slightly obtuse apex, smooth or minutely tuberculate, striate when dry. Peduncle erect or somewhat flexuous; 25–35 cm long, 0.5–0.7 mm in diameter, subterete, 2-ridged above, ridges papillose or not. Inflorescence 2–4-flowered, c.5 × 3 mm, obovoid to ellipsoid. Bracts brown to reddish brown, concolorous, shiny, coriaceous to subcoriaceous, entire, lower ones c.2.5 mm long, ovate-elliptic, rounded at apex, often slightly keeled; intermediate ones c.4 × 2.5 mm, ovate, concave, 5-nerved, rounded and often slightly keeled at apex. Lateral sepals free, 3–3.5 × 0.5 mm, symmetrical, very narrowly elliptic, obtuse at apex; keel narrowed above into a short blunt apiculus, minutely scabrid-ciliate particularly towards apex. Corolla yellow, lobes obovate, 3–3.5 × 2 mm, apex irregularly dentate. Stamens 4 mm long; staminodes 3.5 mm long, bifid, short branches ending in tuft of long yellow hairs. Ovary c.1.5 mm long; style 3–3.5 mm long, trifid. Capsule 3.5–4 mm long. Seeds (mature?) narrowly ellipsoid, 0.6–0.8 × 0.2–0.3 mm, with numerous longitudinal ridges.

Zambia. N: Mbala Dist., Mbala, 11.ii.1957, *Richards* 8149 (K); Kasama Dist., Mungwi, 7.xii.1960, *Robinson* 4164 (K). **Mozambique**. MS: Sussundenga Dist., Tsetserra, Serra Zuira, road to Chimoio, 1800 m, 3.iv.1966, *Torre & Correia* 15,635 (LISC, LMA, LMU, WAG).

Also known from Angola and Congo. Seasonal dambos, 'wet peaty ground'; 1500–1850 m.

Conservation notes: Apparently local, not commonly collected; perhaps Lower Risk near threatened.

In this species the shoot bases tend to be flabellate and, with the dark brown shiny ellipsoid few-flowered inflorescences, give the plant a fairly distinctive appearance. *Xyris asterotricha* is similar but has very distinctive teeth on the keel of the lateral calyx lobes, and *X. capillaris* has much smaller inflorescences.

Material cited by Lisowski (2001) differs from the type in the small stature, smooth pale leaf sheaths, the papillose laminas and smooth peduncle ridges, and in the bracts which are flexuous, dorsally ridged and loosely appressed, with broad pale brown scarious margins. Much of the material comes from high altitudes in the Albertine Rift area and resembles *X. obscura*, but the bracts are not mucronate and the leaf apices are not aristate. It also has a strong superficial resemblance to *X. makuensis* but the lateral sepals of that species have an entire keel, whereas Lisowski's cited specimens have a toothed keel.

14. **Xyris asterotricha** Lock in Kew Bull. **54**: 312, fig.1 (1999). Type: Zimbabwe, Chimanimani Mts, between Uncontoured Peak and Turret Towers, 23.ii.1957, *Goodier* 185 (K holotype, BR, SRGH).

Perennial herb. Leaves up to 42 cm × 0.5 mm; sheaths dark brown, paler towards margins, very glossy, smooth or slightly rugulose above, glabrous; ligule 1.5 mm long, rounded; lamina terete, finely longitudinally ribbed and pitted, glabrous; apex acuminate, symmetrical. Peduncle 30–55 cm long, 0.5–1 mm in diameter, terete, smooth, lightly longitudinally ribbed. Inflorescence obovoid when young, 5–6 × 3 mm, becoming subspherical with age, 8 × 7 mm.

Outer sterile bracts broadly ellipsoid, concave, 3.5 × 2 mm, dark brown, shiny, smooth; apex rounded, mucronate; inner sterile bracts similar but larger, 4 × 2.5 mm; fertile bracts broadly ellipsoid, strongly concave, 5.5 × 4 mm, dark brown, with a scarious margin to 0.5 mm wide. Lateral sepals boat-shaped, lightly curved, narrowly obovate, apex mucronate, the keel with a marginal row of irregularly stellate hairs. Corolla yellow, lobes elliptic, c.4 × 2 mm, apex dentate. Anthers 1.5 mm long; filaments 2 mm long; staminodes much-branched, prominent. Ovary narrowly ellipsoid, 2 × 0.5 mm; style arms 2 mm long; stigmas capitate, flattened. Capsule and seeds not seen.

Zimbabwe. E: Chimanimani Mts, 1700 m, 6.vi.1949, *Wild* 2908 (K, SRGH).

Confined to the Chimanimani Mts in E Zimbabwe, but probably also on the Mozambique side. In high altitude grassland in small rocky valleys and in wet boggy soil; 1700–2100 m.

Conservation notes: Endemic to a small area which is so infertile that it is unlikely to be cultivated; probably Lower Risk near threatened.

The stellate hairs on the sepal keels are matched only by a few specimens of *Xyris congensis*, in particular in the type of its synonym *X. nitida* L.A. Nilsson. However, *X. congensis* is easily distinguished by its inflorescence which elongates with age, and in usually having more than four sterile bracts at the inflorescence base.

15. **Xyris capillaris** Malme in Ark. Bot. **24**(5): 8 (1932). —Lock in F.T.E.A., Xyridaceae: 17 (1999); in Kew Bull. **54**: 313 (1999). —Lisowski *et al.* in F.A.C., Xyridaceae: 63, fig.28 (2001). Types: Angola, between Mt Amaral and Fte. Maria Pia, 2.viii.1905, *Gossweiler* 1829 (BM, LISC); Menongue, I'Kiuna, Caranda, 5.i.1906, *Gossweiler* 2472 (BM, K, LISC); Menongue, Cassuango, R. Kuiriri, 18.x.1906, *Gossweiler* 3246 (BM, K, LISC); Cassuango, R. Kuiriri, 18.x.1906, *Gossweiler* 3267 (B, BM, K, LISC), all syntypes.

Perennial herb forming dense tufts. Leaves up to 13 cm long, markedly distichous at base; sheaths c.2.5 × 0.4 cm near base, pale brown, tinged red when young, smooth, margins and keel ciliate below, otherwise glabrous; ligule whitish, rounded, c.2 mm long; lamina c.10 × 0.05 cm, smooth, ± terete, apex apiculate, symmetrical. Peduncle to 30 cm long and 0.5 mm in diameter, smooth, terete. Inflorescence few-flowered, one flower open at a time, ellipsoid when young, broader when old, c.6 × 3 mm, dark shining brown. Outer sterile bracts broadly ovate, c.4 × 2.2 mm; inner bracts c.4.5 × 3 mm, all concave, crustaceous, brown; fertile bracts similar, c.6 × 4 mm. Lateral sepals 5 × 1 mm, keel denticulate, particularly in lower half. Corolla yellow, tube 3 mm long, lobes broadly ovate, 4 × 3 mm, dentate at apex. Stamens 3 mm long, anthers to 2 mm; staminodes bifurcate, each branch ending in a bunch of long hairs. Ovary oblong, 2 × 0.5 mm; style trifurcate near apex; stigmas much lobed. Seeds ellipsoid, 0.5 × 0.3 mm, with 12–14 longitudinal ridges.

Zambia. W: Mwinilunga Dist., Sinkabolo Dambo, 29.i.1938, *Milne-Redhead* 4391 (K, LISC).

Also in Angola and SW Tanzania. Grassland, 'not very wet'; c.1400 m.

Conservation notes: A poorly-known species with only one collection seen from the Flora Zambesiaca area; Data Deficient, but probably Vulnerable in the Flora area.

Xyris capillaris is perennial with a densely tufted habit and shoots markedly distichous at the base. The inflorescences are much smaller than those of *X. elegantula*.

I have seen only one of the specimens cited by Lisowski in F.A.C. (*Reekmans* 812 from Burundi), but I consider this to be *X. schliebenii*.

Xyris pumila Rendle, also from Angola, is very similar but appears to have broader leaves and darker inflorescence bracts, differences that further collections may show not to be significant. Unfortunately the type (*Welwitsch* 2471 (BM)) is a heavily grazed or burned plant. *Xyris angustifolia* De Wild. & T. Durand, described from Congo (syn. *Xyris exigua* Malme), is superficially similar but has more flattened leaves with rugulose channels on both faces and fimbriate margins to the leaf sheaths. It also occurs in Gabon and Congo (Brazzaville) and appears to be perennial, not annual as in the original description. It has pale brown bracts that are matt, not shining, in contrast to specimens of *X. cf. densa*, also from Congo (Brazzaville), which have smooth (or lightly furrowed, not rugulose or papillose) leaves and shiny brown inflorescence bracts. They also, when burned, have curving old leaf bases as in *X. elegantula*. *Xyris densa* Malme from Congo is also similar and also has smooth leaves; it may be synonymous.

16. **Xyris peteri** Poelln. in Ber. Deutsch. Bot. Ges. **61**: 204 (1944). —Lock in F.T.E.A., Xyridaceae: 19 (1999). Type: Burundi, E of Niakassu, 4.iii.1926, *Peter* 38255 (B holotype).

Perennial herb from slightly swollen base composed of persistent leaf-sheaths and leaf bases. Leaves to 25 × 0.1 cm; sheaths blackish brown, longitudinally ridged, otherwise smooth, shiny, glabrous; ligule rounded, bilobed; lamina linear, smooth or very slightly rugulose, apex acuminate. Peduncle 25–38 cm long, c.1 mm in diameter, glabrous, 2-ridged towards apex. Inflorescence subspherical, c.8 mm in diameter. Sterile bracts very broadly ellipsoid, c.5 × 4 mm, retuse at apex, thinly coriaceous, dark olive-brown; fertile bracts similar. Lateral sepals 6 × 1 mm, apiculate, keel bluntly denticulate. Corolla yellow, lobes broadly ovate, c.4 × 3 mm, dentate at apex. Stamens c.2.5 mm long including 1 mm long anthers. Ovary not seen; style branches trifid at apex; stigmatic surfaces swollen and rounded. Seeds ellipsoid, 0.5 × 0.3 mm, with c.16 longitudinal ridges.

Mozambique. Z: Gurué Dist., Mt Namuli, Muretha Plateau (15°23'50"S; 37°02'49"E), fl. 15.xi.2007, *Harris et al.* 318 (K, LMA).

Also in Rwanda, Burundi, and Tanzania. Upland swamp and grassland; 2250–2800 m.

Conservation notes: Fairly widespread but localised; probably Vulnerable in the Flora area.

Measurements and descriptions of flowers, fruit and seeds are taken from dissections attached to the type.

17. **Xyris subtilis** Lock in Kew Bull. **53**: 892 (1998); in F.T.E.A., Xyridaceae: 17, fig.4 (1999); in Kew Bull. **54**: 313 (1999). Type: Uganda, Masaka Dist., Lake Nabugabo, Bugabo, 28.vii.1971, *Lye & Katende* 6525 (K holotype, EA, MHU). FIGURE 13.4.**2**.

Perennial slender herb. Leaves 8–12 × 0.05–0.1 cm; sheaths rugulose, tapering to apex and expanded at base, with smooth scarious margins which are ciliate particularly towards base; ligule absent; lamina linear, flattened, rugulose; apex solid, acuminate. Inflorescence subspherical, 5–7 mm in diameter at maturity. Peduncle 35–50 cm long, c.0.5 mm in diameter, terete, smooth. Sterile bracts suborbicular, the lowest c.2 mm in diameter; fertile bracts c.4 mm in diameter, thinly coriaceous, 3-nerved, pale olive-brown; apex rounded, often splitting. Lateral sepals arcuate, 4–4.5 mm long, keel bluntly ciliate-denticulate, most markedly so towards middle. Corolla yellow, tube c.5 mm long, lobes elliptic, c.4 × 2 mm. Anthers c.1 mm long; staminodes bifurcate, branches long-ciliate. Ovary ellipsoid, c.3 mm long; style c.3 mm long, 3-branched; stigmas many-lobed. Seeds broadly elliptic, c.0.3 × 0.25 mm, with more than 20 longitudinal ridges.

Zambia. N: Mporokoso Dist., Kasinghi Dambo, 4.ii.1962, *Wright* 353 (K). C: Serenje Dist., Kanona, Mpika road, 8.iv.1932, *St. Clair Thompson* 1284 (K). **Malawi**. N: Mzimba

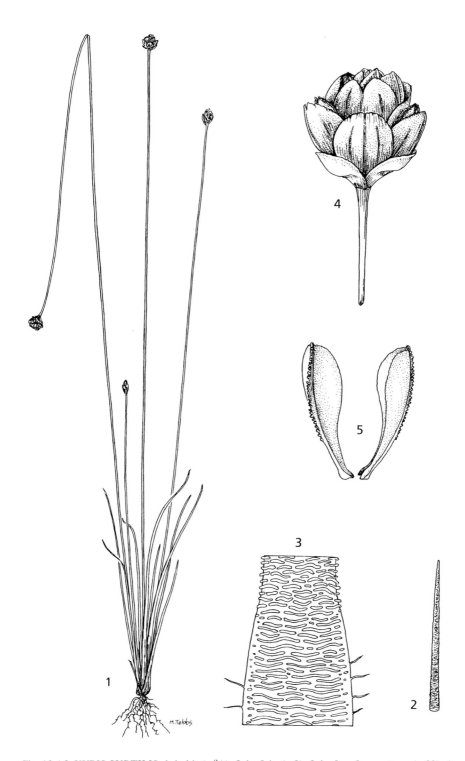

Fig. 13.4.**2**. XYRIS SUBTILIS. 1, habit (× ⅔); 2, leaf tip (× 8); 3, leaf surface pattern (× 20); 4, inflorescence (× 6); 5, lateral sepals (× 8). All from *Lye* 6525. Drawn by Margaret Tebbs. From Flora of Tropical East Africa.

Dist., 9.6 km NW Mzuzu towards Lupaso, 7.iii.1976, *Pawek* 10892 (K). S: Zomba Dist., Zomba Plateau opposite Malosa Saddle, 15.iii.1970, *Brummitt* 9141 (K, SRGH).

Also in Uganda and Tanzania. Moist sandy places and wet hillside flushes; 1350–1900 m.

Conservation notes: Fairly widespread but inconspicuous and probably overlooked; not threatened.

A small slender perennial species with almost spherical blackish brown inflorescences and, in the typical form, rugulose leaves. The specimens listed above probably belong to this taxon, although the leaves are smooth or scarcely rugulose.

18. **Xyris dissimilis** Malme in Ark. Bot. **22**(4): 6 (1928). —Lock in Kew Bull. **54**: 313 (1999). —Lisowski *et al.* in F.A.C., Xyridaceae: 32, fig.13 (2001). Type: Congo, Katanga, Welgelegen, 1.vi.1912, *Corbisier* 636 (BR holotype).

Caespitose perennial herb; shoot bases bulbous, covered by persistent sheaths. Leaves to 24 cm long; sheaths red-brown, shiny when young, smooth; ligule absent; lamina flattened, 24 cm long, 1–1.5 mm wide, glabrous, smooth, margins strongly papillose to very short spreading-hairy, apex solid, curved to one side. Peduncle 40–50 cm long, 0.5–0.7 mm in diameter, terete, smooth or weakly furrowed. Sterile bracts 3.5 × 2 mm, coriaceous, shiny below and cell-patterned above, 5-nerved when viewed from inside; fertile bracts similar but thinner and narrower. Lateral sepals narrowly obovate, c.4 × 1 mm, strongly keeled, keel densely recurved, spinose-hairy except at extreme tip. Corolla yellow, lobes obovate, 3–4 mm long, dentate at tips. Stamens 3–4 mm long; staminodes as long as stamens, bifid, with branches ending in tuft of long yellow hairs. Ovary 1–2 mm long; style trifid. Capsule ellipsoid, c.3.5 × 2 mm. Seeds subspherical, 0.4 × 0.35 mm, dark brown with paler apices, with more than 20 longitudinal ridges.

Zambia. W: Solwezi Dist., Solwezi, 10.iv.1960, *Robinson* 3529 (K).

Also in Congo. In sandy *Brachystegia* woodland; 1350 m.

Conservation notes: Within the Flora area known only from a single specimen; Data Deficient.

The single Zambian specimen, collected at Solwezi not far from the type locality, appears identical to the type. The flattened leaves with rugulose-papillose margins and the lateral sepals with a keel spinulose throughout, but most markedly in the lower two-thirds, are characteristic. Flower details have been taken from Flore de l'Afrique Centrale.

19. **Xyris erubescens** Rendle in Cat. Afr. Pl. Welw. **2**(1): 73 (1899). —N.E. Brown in F.T.A. **8**: 21 (1901). —Malme in Bot. Jahrb. Syst. **48**: 303 (1912). —Lock in Kew Bull. **54**: 314 (1999). —Lisowski *et al.* in F.A.C., Xyridaceae: 38, fig.16 (2001). Type: Angola, Huíla, Nene, by road towards Lopollo, x.1859, *Welwitsch* 2466 (BM holotype, K).

　　Xyris rhodesiana Malme in Bot. Not. **1932**: 12 (1932). —Lock in Kew Bull. **54**: 315 (1999). —Lisowski *et al.* in F.A.C., Xyridaceae: 49, fig.21 (2001). Type: Zimbabwe, Nyanga, near Mt Nyangani, 8.xii.1930, *Fries et al.* 3686 (UPS holotype, BR, SRGH).

　　Xyris theodori Malme in Bot. Not. **1932**: 13 (1932). Type: Zimbabwe, Nyanga, near Mt Nyangani, 8.xii.1930, *Fries et al.* 3676 (LD holotype, BM).

Bulbous-based perennial herb, flowering as or before the leaves develop. Leaves poorly known, barely developed at time of flowering. Lamina ± terete, to 26 cm long (probably more when fully developed), c.1 mm in diameter, weakly longitudinally ridged, otherwise smooth; apex long-aristate, hyaline. Sheaths coriaceous to scarious, long-acuminate from a broad base, blackish brown, sometimes rugulose and with hairy margins at extreme base, otherwise smooth. Peduncle subquadrangular, 14–40 cm tall, c.1 mm in diameter, glabrous. Inflorescence broadly obovoid to subglobose, 4–6 mm in diameter. Outer bracts oblong, 4–6 × 3–4 mm, yellowish brown with pale entire margins; fertile bracts orbicular, concave, shortly apiculate or retuse, 7–9-nerved. Lateral sepals 5 × 1 mm, falcate, strongly keeled, keel minutely hispidulous from middle to apex. Staminodes 2-armed. Other floral details not known. Seeds elongate, 1.6–1.8 × 0.3–0.4 mm, smooth or faceted.

Zambia. N: Mporokoso, 130 km NW of Kasama, 17.x.1967, *Simon & Williamson* 1085 (K, LISC). W: Ndola Dist., between Broken Hill zinc mine and Bwana Mkubwa, x.1906, *Allen* 348 (K). **Zimbabwe**. C: Marondera Dist., Marondera (Marandellas), 5.xii.1939, *Hopkins* 7476 (SRGH). E: Chimanimani Dist., Chimanimani Mts, upper Bundi Plateau, 31.xii.1957, *Goodier & Phipps* 243 (K, SRGH). **Malawi**. N: Rumphi Dist., Nyika, Chelinda Bridge, 21.xii.1969, *Pawek* 3275 (K). **Mozambique**. MS: Sussundenga Dist., Rotanda, between rio Mussapa and frontier, 19.xi.1965, *Torre & Correia* 13165 (BR, LISC, LMA, LMU, WAG).

Also in Angola and Congo, and probably Tanzania. Moist places in montane grassland, dambos; 1400–1850 m.

Conservation notes: Quite widespread, but with taxonomic problems (see below); not threatened.

The type is poor and lacks leaves. As I interpret this taxon, it has leaf sheaths that are smooth and shiny, without ciliate margins at the base, lateral sepals with rather broad-based, flattened teeth mainly near the middle, and leaves which are not rugulose. There is some variation in inflorescence size, perhaps caused by water availability at the time of flowering, but also suggesting that more than one taxon may be involved.

Lisowski places *Xyris sphaerocephala* Malme in the synonymy of this species. I cannot agree; all the specimens of *X. sphaerocephala* come from the Kinshasa region of Congo; the upper parts of the leaf sheaths are strongly rugulose, and the inflorescence bracts are chestnut-brown rather than yellowish.

Allen 348 (K) was named *X. friesii* by Malme in 1925; in 1933 he renamed it "*Xyris sphaerocephala Malme*", but suggested it may be a juvenile specimen of *X. sphaerocephala* and also resembled *X. erubescens* Rendle.

20. **Xyris capnoides** Malme in Ark. Bot. **24**(5): 4 (1932). —Lock in Kew Bull. **54**: 315 (1999). Type: Angola, Huíla, Kuvango (Forte Princesa Amélia), 22.x.1905, *Gossweiler* 2127 (BM ?holotype, LISC).

Perennial herb from a bulbous base. Leaves weakly developed at flowering, subterete to weakly flattened, 9–13 × 0.1 cm, subulate at apex, smooth, distinctly longitudinally striate; sheaths c.3 cm long, with a distinct ligule, wider than lamina above, widened below and usually glabrous (some with ciliate margins) but minutely papillose or transversely rugulose, chestnut- or red-brown. Peduncle 20–28 cm tall, c.1 mm in diameter, often flexuous, subterete, smooth, surrounded by basal sheath 7–9 cm long. Inflorescence few-flowered, 6–7 × 4–5 mm, obovoid. Inflorescence bracts subcoriaceous, margins ± scarious, entire, pale brown, concolorous, shiny, lower ones ovate, c.4 × 3 mm, rounded at apex and sometimes mucronulate, indistinctly keeled below apex; intermediate ones conchoidal or convex, ovate, c.5 × 3.5 mm, rounded at apex, unkeeled. Lateral sepals free, subequilateral, lanceolate, c.4 × 0.8 mm, acute, smoky, with a narrow keel, usually denticulate or cilate-scabrid from base to apex. Flowers, fruits and seeds unknown.

Zambia. W: Mwinilunga Dist., SW of Dobeka bridge, 11.x.1937, *Milne-Redhead* 2693 (in part) (K).

Also in Angola. In dambos; c.1400 m.

Conservation notes: Very local distribution, and in a heavily disturbed habitat; Data Deficient.

21. **Xyris angularis** N.E. Br. in F.T.A. **8**: 22 (1901). —Lock in F.T.E.A., Xyridaceae: 13 (1999); in Kew Bull. **54**: 315 (1999). Type: Nigeria, Nupe, *Barter* s.n. (K holotype).

Xyris vanderystii Malme in Ark. Bot. **22**(4): 3 (1928). —Lisowski *et al.* in F.A.C., Xyridaceae: 21, fig.8 (2001). Types: Congo, Kalchaka, xii.1913, *Vanderyst* 2840 (BR lectotype, designated by Lewis 1972); Kimbambi, i.1925, *Vanderyst* 14745; Atene, 1914, *Vanderyst* 3144; Atene, i.1914, *Vanderyst* 3431 bit; N'Dembo, 5.i.1907, *Vanderyst* s.n. (all BR syntypes).

Xyris decipiens N.E. Br. var. *vanderystii* (Malme) Malme in Ark. Bot. **24**(5): 2 (1932).

Xyris decipiens sensu Hepper in F.W.T.A. ed.2, **3**(1): 55 (1967), non N.E. Br.

Perennial herb forming loose clumps. Leaves 8–15(40) × 0.3–0.4 cm; sheaths lightly furrowed, brown, glabrous, with narrow scarious margins; ligule acute, c.1 mm long; lamina linear, glabrous, smooth or minutely rugulose on one surface; apex acute, straight or slightly incurved, serrulate, sometimes with group of small finger-like hairs at extreme tip. Peduncle 20–50(85) cm long, 1–1.5 mm in diameter, terete with 4–6 longitudinal ridges, glabrous. Inflorescence subspherical to broadly ellipsoid, up to 8 × 8 mm at maturity, reddish brown. Sterile bracts broadly ovate, c.5 × 4 mm, coriaceous, reddish brown with very distinct greyer central area, the lowest smaller; margins lacinate-ciliate, sometimes with distinct apical tuft of tangled hairs, later glabrous. Lateral sepals arcuate, 4.5 × 1 mm, keel ciliate-dentate, sometimes with apical tuft of hairs. Corolla yellow, tube c.3 mm. long; lobes oblong, c.3.5 × 1.5 mm, dentate at apex. Stamens c.1 mm. long; anthers 0.5 mm long, lobes spreading below; staminodes shortly bifid with small bunches of hairs. Ovary oblong, c.2 × 0.5 mm; style 4 mm long, 3-branched in upper third; stigmas many-lobed. Seeds narrowly ellipsoid, c.1 × 0.25 mm, with numerous irregular longitudinal ridges or reticulations.

Zambia. B: Mongu Dist., 40 km NE of Mongu, 12.xii.1965, *Robinson* 6727 (EA, K, LISC, SRGH). N: Chinsali Dist., Chipoma Falls, Jubu R., 22.5 km S of Chinsali, 26.x.1967, *Simon & Williamson* 1208 (SRGH). W: Mwinilunga Dist., source of Zambezi R., 13.xii.1963, *Robinson* 5994 (K, SRGH). **Zimbabwe**. E: Chimanimani Dist., upper Haroni, 31.i.1957, *Phipps* 240 (K, SRGH).

Also in Benin, Nigeria, Cameroon, Central African Republic, Congo (Brazzaville), Congo, Burundi and Angola. Seasonally or permanently wet sandy grassland or woodland, swamp edges; extends well into the forest zone outside the flora area; 1000–1300 m.

Conservation notes: A widespread species, although never common; not threatened.

A small perennial species, distinctive in its strongly patterned inflorescence bracts. The leaves also often have a distinctive speckled appearance. Material from Zimbabwe lacks the basal parts but the bracts are sufficiently distinctive for identification; further material would be welcome.

The true *Xyris decipiens* is very similar but has much larger inflorescences. It is mainly a West African coastal species but is also found inland. The type is the most southerly specimen I have seen.

Lisowski (in Flora of Guinea 2009) has suggested that *X. decipiens* is synonymous with *X. macrocephala* Vahl from South America. Comparison with South American material at K and BR (there are several South American specimens at Kew from the 1840s) suggests that some specimens named as *X macrocephala* are indeed very similar. A definite conclusion must await more detailed comparison and possibly molecular studies. Similar distributions are found in other taxa but it difficult to know if they are due to trans-ocean dispersal or to early introduction.

22. **Xyris anceps** Lam., Tab. Encycl. **1**: 132 (1791). —N.E. Brown in F.T.A. **8**: 12 (1901). —Malme in Bot. Jahrb. Syst. **48**: 307 (1912). —Hepper in Kew Bull. **21**: 419 (1968). —Lewis & Obermeyer in F.S.A. **4**(2): 2 (1985). —Lock in F.T.E.A., Xyridaceae: 12 (1999); in Kew Bull. **54**: 316 (1999). —Lisowski *et al.* in F.A.C., Xyridaceae: 96, fig.41 (2001). Type: Madagascar, *Commerson* s.n. (P holotype, BM).

Var. **anceps**. —Lock in Kew Bull. **53**: 890 (1998).

Perennial herb, perhaps sometimes annual, forming large clumps. Leaves linear, 20–40 × 0.4–0.5 cm, glabrous, brownish or reddish when dry, tapering gradually to asymmetric apex. Peduncle up to 100 cm long, 1.5–3 mm in diameter, glabrous and longitudinally ridged when dry, somewhat flattened and markedly 2-winged, particularly towards apex. Inflorescence broadly ovoid, later ovoid, up to 15 × 11 mm. Sterile bracts 4, suborbicular, 4–5 mm long and broad, bifid at apex, pale golden brown; fertile bracts 30–40, ovate to elliptic, c.6 × 4 mm, slightly bifid, with small rhombic slightly discolourous patch near apex. Lateral sepals 5 mm. long, curved, keel prominent, entire, often with a submarginal reddish line near apex. Corolla yellow, tube c.3 mm long, lobes oblong, c.4 × 2 mm. Stamens 2 mm long; anthers c.1 mm long; staminodes branched, each branch fimbriate at apex. Ovary obovoid, c.2 × 1 mm; style branches 1.5 mm long; stigma a dense tuft of short branchlets at apex. Capsule ellipsoid, 4 × 2 mm. Seeds numerous, broadly ellipsoid, with 12–13 longitudinal ridges, 0.5 × 0.4 mm.

Mozambique. Z: Pebane Dist., 75 km from Pebane to Mualama, 20 m, 10.iii.1966, *Torre & Correia* 15159 (LISC, LMA, LMU, WAG). GI: Bilene Dist., São Martinho, 30.xi.1955, *Gonçalves-Sanches* 22 (LISC). M: Maputo Dist., between Manhiça and Bilene, Páti (Pate) lagoon, 25.iii.1954, *Barbosa & Balsinhas* 5462 (BM, LISC).

Also along the East African coast from S Kenya to KwaZulu-Natal, in Madagascar and, more rarely, the West African coast from Nigeria to Angola. Coastal grassy swamps on sand; 0–60 m.

Conservation notes: Although coastal swamps are susceptible to drainage for agriculture, aquaculture and tourism, its area of occurrence is very large; not threatened.

The flowers of var. *anceps* are hardly ever present in herbarium material and the measurements of floral parts given here should be regarded as approximate.

Var. **minima** (Steud.) Lock in Kew Bull. **53**: 890 (1998); in F.T.E.A., Xyridaceae: 13 (1999); in Kew Bull. **54**: 316 (1999). Type: Guinea Republic, near Conakry, Los Is., Crawford, *Jardin* B120 (S isotype).

Xyris minima Steud., Syn. Pl. Glumac. **2**: 288 (1855).
Xyris perrottetii Steud., Syn. Pl. Glumac. **2**: 319 (1855). Type: Senegal, n.d., *Perrottet* 809 (STU).
Xyris humilis var. *minima* (Steud.) T. Durand & Schinz, Consp. Fl. Afr. **5**: 420 (1895).

Annual herb. Leaves markedly distichous, linear, 8–15(25) × 0.3–0.5 cm, glabrous, drying greenish, parallel-sided, tapering rather abruptly to an asymmetrical obtuse apex. Peduncle (14)20–45 cm long, 1–1.5 mm in diameter, glabrous, hollow, terete but with 2 longitudinal wings. Inflorescence subspherical, 7–8 × 7–8 mm. Sterile bracts 4, broadly obovate, c.3 × 2 mm, pale golden-brown with a darker median line and subterminal narrowly ovate greyish dorsal mark; fertile bracts suborbicular, apiculate, very convex, c.3 × 3 mm, golden-brown with dull brown to grey subapical triangular area c.1 mm long. Lateral sepals 3 mm long, curved, golden-brown, with slight entire-margined keel. Corolla yellow, tube c.1.5 mm long, lobes oblong, c.2 × 1.5 mm. Stamens c.1.5 mm long; anthers 1 mm long; staminodes shortly bifurcate, branches bearing tufts of hairs. Ovary ellipsoid, 1.5–1 mm; style 3 mm long, trifid in upper third; stigmas a dense mass of short branches. Capsule broadly ellipsoid, c.3 × 1.5 mm. Seeds numerous, ellipsoid, striate with 12–13 longitudinal ridges, c.0.3 × 0.2 mm.

Mozambique. N: Marrupa Dist., 25 km from Marrupa to Lichinga, fl. 8.viii.1981, *Jansen et al.* 144 (EA). Z: Nicoadala Dist., Nicoadala, viii.1887, *Scott* s.n. (K).

Found in West Africa from Senegal to Gabon and in East African coastal regions. Coastal and inland seasonal swamps and wet places, often a weed of rice fields; 20–700 m.

Conservation notes: Widespread and often weedy; more widespread in Africa than var. *anceps* but apparently rare in south-central Africa. Not threatened at global level, although perhaps Vulnerable in the Flora area.

Var. *anceps* and var. *minima* have generally been considered to be the same. They are, however, usually clearly distinct in the herbarium and apparently also in the field, and it seems wrong not to recognize this fact, although intermediates exist.

23. **Xyris imitatrix** Malme in Ark. Bot. **22**(4): 2 (1928). —Lock in Kew Bull. **54**: 316 (1999). —Lisowski *et al.* in F.A.C., Xyridaceae: 17, fig.6 (2001). Types: Congo, Luebo-Kasai region, Kinshasa (Leopoldville), x.1915, *Achten* 369B (BR lectotype, designated by Lewis 1972); *Achten* 369a; Leopoldville, x.1915, *Achten* 403a; Boko, v.1913, *Vanderyst* s.n.; between Dembo and Kisantu, x.1900, *Gillet* 1596 (all BR syntypes).

Perennial herb forming loose clumps. Leaves linear, 10–30 × 0.4–0.5 cm, falcate-incurved at acute apex, striate when dry, smooth or minutely papillose, glabrous on the faces, with long weak hairs on margins. Sheaths $^1/_3$ to $^1/_4$ length of lamina. Peduncle 30–65 cm long, c.1 mm in diameter, hollow, with 6–8 longitudinal ridges, brownish and shiny towards base. Inflorescence few-flowered, obovoid to ellipsoid, 8–10(12) × 4–5(6) mm. Inflorescence bracts entire, unkeeled, rounded at apex, lower ones 3–4 mm long, ovate, with distinct dorsal mark; middle ones obovate to suborbicular, 5–6 × 4–5 mm, convex, chestnut-brown, semi-opaque, dorsal area 2–2.5 mm long, grey-green. Lateral sepals c.3 × 0.5 mm, free, subhyaline, lanceolate or narrowed from slightly wider base, acute, wing of keel narrow, sparsely lacerate-ciliate. Capsule c.3.5 mm long; pericarp very thin, whitish, placentae scarcely confluent at base. Seeds oblong-cylindric, c.1 mm long.

Zambia. B: Zambezi Dist., Zambezi, iii.1960, *Pinhey* 13 (K, SRGH). W: Mwinilunga Dist., 6 km N of Kalene Hill, 12.xii.1963, *Robinson* 5933 (K, SRGH).

Also in Gabon, Congo and Congo (Brazzaville). Permanently wet dambos, but outside the Flora area it extends far into the forest zone, occurring in natural swampy clearings (ésobés); 1000–1300 m.

Conservation notes: A relatively widespread species, although only two collections from the Flora area; best treated as Data Deficient.

Xyris imitatrix is unique among African species in its pilose leaf margins and in the leaves which are neatly arranged in a single plane. It is very distinctive and hard to confuse with any other species.

24. **Xyris capensis** Thunb., Prodr. Pl. Cap.: 12 (1794). —Rendle in Cat. Welw. Afr. Pl. **2**(1): 68 (1899). —N.E. Brown in F.T.A. **8**: 13 (1901). —Malme in Bot. Jahrb. Syst. **48**: 305 (1912); in Svensk Bot. Tidskr. **6**: 557 (1912); in Bot. Not. **1932**: 14 (1932). —Hepper in F.W.T.A. ed.2, **3**: 54 (1968). —Lewis & Obermeyer in F.S.A. **4**(2): 2 (1985). —Lock in F.T.E.A., Xyridaceae: 16 (1999); in Kew Bull. **54**: 316 (1999). — Lisowski *et al.* in F.A.C., Xyridaceae: 65, fig.29a (2001). Type: South Africa, Cape, near Verkeerde valley, 1772–3, *Thunberg* 1267 (UPS holotype).

Xyris zombana N.E. Br. in F.T.A. **8**: 13 (1901). Type: Malawi, Mt Zomba, xii.1896, *Whyte* s.n. (K holotype).

Annual or sometimes perennial herb, forming loose clumps. Leaves flabellate, to 40 × 0.2 cm; sheath gradually dilated to base, glabrous, smooth, green, brown at base, margin hyaline; ligule absent; lamina linear, glabrous; apex solid, paler, acute, slightly incurved. Peduncle 30–50(75) cm long, c.0.2 cm in diameter, ± terete with low longitudinal ridges. Inflorescence subspherical, becoming depressed-spherical with age, dark brown. Sterile bracts broadly ellipsoid to broadly obovate, rounded, the outer often split at apex, brown, darker towards middle and apex; fertile bracts similar but paler and thinner, strongly concave, midrib keeled towards retuse apex and apiculate. Lateral sepals boat-shaped, curved, glabrous, pale brown, keel entire, dark brown towards the acute or acuminate apex. Corolla yellow, lobes broadly obovate, c.3.5 × 2.5 mm; tube c.4.5 mm long. Anthers c.1 mm long; filaments c.0.5 mm long; staminodes bifurcate, much-branched. Ovary obovoid, 2.5 × 1.5 mm; style 2 mm long, branches c.1 mm long, much branched; stigmas aggregated into a single mass. Seeds ellipsoid, 0.3 × 0.2 mm, with 12–16 longitudinal ridges; cross-ridges also clear.

Botswana. N: Santantadibo R. at junction with cross-channel from Mboroga, 11.i.1978, *P.A. Smith* 2188 (K, SRGH). **Zambia**. B: Mongu Dist., Sofula, 25.viii.1960, *Robinson* 5460 (BR, K, SRGH). N: Chinsali Dist., Shiwa Ngandu, far side of Lake Young, 19.i.1959, *Richards* 10742 (K). W: Mwinilunga Dist., Kalenda Plain, 9.x.1937, *Milne-Redhead* 2686 (K). C: Kabwe Dist., Mwapula R., Mwiya village, 15.vi.1996, *Bingham* 11045 (K). S: Kalomo Dist., Machili, 4.vii.1960, *Fanshawe* F5770 (K, SRGH). **Zimbabwe**. W: Matobo Dist., Farm Besna Kobila, v.1957, *Miller* 4386 (K, SRGH). C: Prince Edward Dam, 10 km S of Harare, 14.x.1955, *Drummond* 4894 (K, LISC, SRGH). E: Nyanga Dist., 10 km N of Troutbeck, Gairesi Ranch, 15.xi.1956, *Robinson* 1904 (K, SRGH). S: Chivi Dist., near Madzivire Dip, 7 km E of Runde R. Bridge, 3.v.1962, *Drummond* 7882 (K, SRGH). **Malawi**. N: Mzimba Dist., Katoto, 5 km W of Mzuzu, 3.viii.1973, *Pawek* 7291 (K, SRGH). C: Kasungu Dist., Kasungu Nat. Park, 6.ix.1972, *Pawek* 5678 (K, SRGH).

Widespread in tropical Africa, mainly in upland areas. Upland bogs and marshes; 400–1800 m.

Conservation notes: A widespread taxon; not threatened.

This is a highly variable and widespread taxon which may be a complex of several. There is need for a comprehensive revision throughout its range, which is usually understood to include India and southeast Asia. Within south-central Africa, virtually all specimens appear to be annual, with distichous leaf bases and subspherical dark brown inflorescences. The scape is terete or lightly ridged. By contrast, in East Africa there is a well-marked perennial form that has been collected mainly from the Uluguru Mts in Tanzania; no similar plants have been seen from further south.

Another group, represented in East Africa by *Haarer* 2074, *Kahurananga et al.* 2755, *Milne-Redhead & Taylor* 10849, and *Drummond & Hemsley* 4688, includes clearly annual plants with strongly distichous leaves with purple bases and rather obtriangular inflorescences. Such material has often been confused with *X. straminea* and named as *X. multicaulis*, a synonym of the former. While distinct as a group, these specimens are connected to 'normal' *X. capensis* by intermediates.

The type of *X. zombana* is a robust plant which can be matched to other material from upland areas elsewhere in Africa. It is connected to typical *X. capensis* by intermediates.

Several varieties of *X. capensis* have been described and some of these are recognisable in south-central Africa, although they do not seem any more distinct than other forms. Specimens from the Okavango Swamps (Botswana), such as *P.A. Smith* 272 (SRGH), 349 (SRGH) & 2774 (SRGH), named as *X. capensis* var. *microcephala* Malme, are here regarded as *X. huillensis*.

25. **Xyris porcata** Lock in Kew Bull. **54**: 317, fig.2 (1999). Type: Zambia, Mbala Dist., Lake Chila, 29.xii.1956, *Richards* 7445 (K holotype).

Annual or short-lived perennial herb forming small tussocks, sometimes with a thin vertical rhizome to 7 cm long, perhaps developed in response to rising water levels. Leaf sheaths to 10 cm long, pale brown, smooth, glabrous, margins membranous, very pale, glabrous, entire. Lamina 20–23 × 0.2–0.3 cm, linear, flat, smooth, acute and slightly asymmetric at apex. Scapes up to c.45 cm tall, 1 mm in diameter, terete, 2-ridged at least in upper part. Inflorescence ellipsoid, to 13 × 5 mm, more usually c.9 × 5 mm. Lowest pair of sterile bracts c.5 × 3 mm, ovate, strongly keeled dorsally, acute at apex, pale brown with translucent margins; upper pair of sterile bracts c.5.5 × 4 mm, broadly ovate, strongly keeled dorsally, acute at apex, pale brown but darker towards middle and apex; fertile bracts c.5.5 × 4 mm, very broadly ovate, keeled towards apex, chestnut-brown, darker above and almost black near apex, margins minutely spinulose towards apex. Lateral sepals 4.5 × 1.5 mm, arcuate, pale brown, darker on spinulose keel. Corolla yellow, lobes elliptic, c.4 × 2 mm. Anthers c.2 mm long; staminodes much branched, branches filamentous. Ovary ellipsoid, c.3 × 1.5 mm; style arms c.2 mm long, capitate. Capsule not seen. Seeds ellipsoid, 0.5 × 0.25 mm, with c.10 longitudinal ridges.

Zambia. N: Mbala Dist., Lake Chila, fl.& fr. 4.i.1952, *Richards* 230 (K); Samfya Dist., Samfya, fl. 2.x.1971, *Fanshawe* 342 (EA, K, SRGH).

Known only from Zambia. Lakeshores; 1500–1700 m.

Conservation notes: Endemic to a restricted area in N Zambia, where it does not appear to be common; probably Vulnerable.

A distinctive species with strongly ridged dorsal surfaces of the inflorescence bracts, to which the specific name refers.

26. **Xyris makuensis** N.E. Br. in F.T.A. **8**: 17 (1901). —Malme in Bot. Jahrb. Syst. **48**: 304 (1912). —Lock in F.T.E.A., Xyridaceae: 15 (1999); in Kew Bull. **54**: 318 (1999). Type: Mozambique, Mt Namuli, 'Makua Country', 1887, *Last* s.n. (K holotype, B). Although other specimens are cited as syntypes, Brown's annotation on the sheet and the name suggests he regarded Last's specimen as the type.

Perennial herb forming clumps; rhizomes sometimes present, upright or ascending. Leaves 3–18 × 0.1 cm; sheaths pale brown, smooth, glabrous, gradually tapered to apex; ligule absent; lamina flattened, glabrous, smooth or with minutely denticulate margins; tip solid, acute, sometimes incurved. Peduncle 10–26 cm long, smooth, terete, weakly 2-ridged above. Inflorescence elliptic, 5–6 mm long, narrowly ovoid; bracts dark brown, midrib almost black, forming a keel towards apex, excurrent as a mucro. Lowest sterile inflorescence bracts elliptic, c.2.5 × 1.5 mm; upper sterile bracts elliptic, 3.5 × 2 mm; fertile bracts 4.5 × 2.5 mm. Lateral sepals lightly curved, c.5.2 × 1 mm, pale brown with very dark keel, particularly towards apex; keel entire. Corolla lobes suborbicular, c.4.5 mm. in diameter, apex (?always) dentate. Anthers 1.5 mm long; filaments c.1 mm long; staminodes much-branched, forming dense tufts of hairs. Ovary obovoid, c.3 mm long; style 5.5 mm long in total, 3-branched, branches 2.5 mm long; stigmas peltate, many-lobed. Seeds ellipsoid, 0.6–0.7 mm long, with c.12 longitudinal ridges.

Malawi. S: Mulanje Dist., Mt Mulanje, above top of Little Ruo waterfall, 28.vii.1970, *Brummitt* 12325 (K, LISC, SRGH). **Mozambique**. Z: Gurué Dist., W of Namuli Peaks, 5.i.1968, *Torre & Correia* 16930 (LISC).

Confined to S Malawi and NC Mozambique; the record from Tanzania (Lock 1999) is almost certainly incorrect. Upland grassland and swamp; 1600–2000 m.

Conservation notes: Apparently endemic to Mts Namuli and Mulanje, although appearing to be common there; Lower Risk near threatened.

The very dark brown elongated inflorescences and strongly keeled bracts are distinctive.

27. **Xyris huillensis** Rendle in Cat. Afr. Pl. Welw. **2**(2): 71 (1899). —N.E. Brown in F.T.A. **8**: 18 (1901). —Lock in F.T.E.A., Xyridaceae: 15 (1999); in Kew Bull. **54**: 306 (1999). Types: Angola, Huíla, Empalanca, ii.1860, *Welwitsch* 2469 (BM syntype, K, M, PRE); Humpata, between Nene and Humpata, near rio Quipumpunhine, v.1860, *Welwitsch* 2472 (BM syntype, K).

Xyris capensis Thunb. var. *medullosa* N.E. Br. in F.T.A. **8**: 14 (1901). —Lisowski *et al.* in F.A.C., Xyridaceae: 67, fig.29b (2001). Type: Tanzania, ii.1887, *Hannington* s.n. (K holotype).

Small tufted perennial herb. Leaves 5–8(11) × 0.1 cm; sheaths brown, smooth, glabrous except for ciliate margins below; ligule not seen; lamina linear, smooth, glabrous; apex acute, solid, smooth. Peduncle 20–30 cm long, terete, glabrous, smooth, weakly grooved longitudinally; sheaths smooth, shorter than leaves. Inflorescence ellipsoid, dark brown, 4–5 mm long. Outer sterile bracts obovate, 3.5 × 1.8 mm, brown, apex emarginate, nerve strong but poorly defined, excurrent as a mucro; inner sterile bracts similar but nerve less prominent and apiculus lacking, apex often splitting but probably not truly laciniate; fertile bracts similar but thinner and more concave. Lateral sepals boat-shaped, 3.5 × 0.8 mm, pale brown, darker on entire glabrous keel. Corolla yellow, tube 3 mm long, lobes oblong, 2 × 1 mm. Stamens 1 mm long including 0.5 mm anthers; staminodes bifid, each branch with a bunch of hairs. Ovary obovoid, 2 × 1 mm; style 3 mm long, the upper × trifid; stigmas bi- or trilobed. Capsule obovoid, 6 mm long. Seeds ellipsoid, 0.5 × 0.3 mm, longitudinally ridged.

Botswana. N: Okavango Delta, Mboma stream, 18.i.1973, *P.A. Smith* 349 (K, SRGH). **Zambia**. B: Kaoma Dist., near Mankoya Resthouse, 20.xi.1959, *Drummond & Cookson* 6673 (K, SRGH). N: Mbala Dist., stream above Chilongowelo Farm, 18.xii.1951, *Richards* 91 (K). W: Mwinilunga Dist., by Kewumbo R., 15.xii.1937, *Milne-Redhead* 3683 (K). **Zimbabwe**. C: Marondera Dist., Looe, 22.xii.1948, *Wild* 2714 (K, SRGH). E: Mutare Dist., Odzani R. valley, 1914, *Teague* 410 (BOL, K). **Malawi**. N: Mzimba Dist., Katoto, 5 km W of Mzuzu, 3.viii.1973, *Pawek* 7277 (EA, K, MAL, MO, SRGH).

Also known from Tanzania and Angola. In upland swamps and bogs; 950–1400 m.

Conservation notes: Widely distributed but probably much overlooked; not threatened.

A small and rather nondescript species, doubtless frequently overlooked as a small perennial form of *Xyris capensis*. The obconical form of the inflorescences and the smooth brownish bracts help in identification; the bases of the leaf sheaths often have a fringe of cilia.

Lewis, when labelling herbarium sheets, often applied the name *X. capensis* var. *medullosa* to small plants which are here treated as *X. rubella* and *X. schliebenii*. In Flora of Cameroun he placed this as a synonym of *X. capensis*; in Flora of Southern Africa he does not mention it but treated *X. rubella* as a synonym of *X. capensis*.

Haarer 1660 (K) from Tanzania was seen by Malme, who annotated it as '*Species nova e stirpe X. capensis*'. He did not connect it with *X. capensis* var. *medullosa*, nor with *X. huillensis*. Malme also identified *Teague* 389 & 410 (both K) from Zimbabwe as *X. capensis* var. *medullosa* (= var. *microcephala* Malme). I regard these as *X. huillensis*.

28. **Xyris rhodolepis** (Malme) Lock in Kew Bull. **53**: 886 (1998); in F.T.E.A., Xyridaceae: 14 (1999); in Kew Bull. **54**: 320 (1999). —Lisowski *et al.* in F.A.C., Xyridaceae: 85, fig.35 (2001). Types: Angola, Cuanavale, rio Domba, 31.vii.1906, *Gossweiler* 2817 (BM syntype, K); I'Chancambe, by rio Cuanavale, 12.viii.1906, *Gossweiler* 2644 (BM syntype, K, LISC); between Mt Amaral and Fte. Maria Pia, 2.viii.1905, *Gossweiler* 1829A (BM syntype).

   Xyris humpatensis N.E. Br. var. *rhodolepis* Malme in Ark. Bot. **24**(5): 7 (1932).

Small tufted perennial herb. Leaves 3–5(8) × 0.1 cm; sheaths brown, rugulose, appressed-rufous-pilose at extreme base, cilate on margins and irregularly ciliate on 2 dorsal ridges; lamina linear, flattened, rugulose throughout; apex acute, solid, smooth. Peduncle 14–30 cm long, terete, smooth, glabrous. Inflorescence obovoid, 5–6 mm long, dark red-brown. Outer sterile bracts elliptic, c.4 × 2 mm, apex obtuse with single strong nerve; inner bracts similar but thinner and 3-nerved; fertile bracts similar but broader and thinner, often papillose towards apex. Lateral sepals c.4 × 1 mm, strongly keeled, keel irregularly ciliate-toothed, teeth below middle often directed downwards, red-brown, keel darker, apex apiculate. Corolla lobes yellow, broadly obovate, c.3.5 × 3 mm. Anthers c.1 mm long; filaments c.0.5 mm long; staminodes a bunch of hairs. Ovary ellipsoid (from F.A.C.), 1.5 mm long; style trifid. Seeds ellipsoid to ovoid, 0.5 × 0.25 mm, with c.12 longitudinal ridges.

Zambia. B: Senanga Dist., fringe of Lui R., 15.vii.1964, *Verboom* 1755 (K). W: Kalulushi Dist., Chingola, 26.viii.1954, *Fanshawe* F1498 (EA, K, SRGH). **Zimbabwe**. C: Marondera Dist., Grasslands Research Station, 19.viii.1963, *Boughey* 12231 (K). **Malawi**. N: Mzimba Dist., 5 km W of Mzuzu, Katoto, 3.viii.1973, *Pawek* 7290 (K). **Mozambique**. N: Lago Dist., 100 km N of Maniamba, 13.ix.1934, *Torre* 226 (LISC).

Also in Tanzania, Congo and Angola. Wet grasslands, often near rivers; 1000–1500 m.

Conservation notes: Fairly widely distributed but probably much overlooked; not threatened.

At first sight this is a small and rather nondescript perennial taxon, but the reddish papillose bracts are very different from the smooth yellowish ones of *X. humpatensis* and from all other African species.

Two specimens from Angola, *Barros Machado* ANG.VI.54-128 (LISC) from Alto Cuílo, rio Cavuemba and *Gossweiler* 11847 (LISC) from Lunda, Xa-Sengue, Caiango, have markedly elongated outer inflorescence bracts but are otherwise similar to normal *X. rhodolepis*.

29. **Xyris welwitschii** Rendle in Cat. Afr. Pl. Welw. **2**(2): 68 (1899). —N.E. Brown in F.T.A. **8**: 15 (1901). —Lock in Kew Bull. **53**: 891 (1998); in Kew Bull. **54**: 306 (1999). Type: Angola, Huíla, between Lopollo and Monino, iv.1860, *Welwitsch* 2465 (BM lectotype selected by Lock 1998, K, M).

Annual herb; stems solitary, to 35 cm tall. Leaf sheaths green-brown, smooth; ligule absent; lamina flattened, 5–12 × 0.15–0.2 cm; surface smooth, glabrous, margins papillose; apex aristate. Peduncle 7–30 cm long, 0.5–0.7 mm in diameter, glabrous, smooth, flattened above, with pale ridges on angles and surface. Inflorescence subspherical at maturity, yellow to pale brown; lowest (sterile) bracts 3.5 × 2.5 mm, broadly ovate, concave, 3-veined, yellow-brown; second pair of bracts similar but larger, 4.5 × 3 mm; fertile bracts similar but larger, 5.5 × 3 mm. Lateral sepals 5.5 × 1.5 mm, narrowly obovate, pale yellow; keel entire. Flowers not seen. Capsule 4 × 2.5 mm, obovoid. Seeds broadly ellipsoid, 0.4 × 0.3 mm, blackish, with 10–12 longitudinal ridges linked by distinct cross-ridges.

Zambia. N: Mpika Dist., Serenge–Mpika road, 5.iv.1961, *Richards* 15008 (BR, K).

In the sense used here, known only from Angola and Zambia. Damp sandy soils, among rocks; 1200 m.

Conservation notes: Known from only a few widely scattered specimens; probably Vulnerable in the Flora area.

As accepted here, *X. welwitschii* is an annual species with subspherical yellowish inflorescences. These are larger than those of other annual species, with the exception of *X. capensis* from which it differs most obviously in the papillose leaf margins. For discussion of typification and the use of the name see Lock (1998).

30. **Xyris straminea** L.A. Nilsson in Öfvers. Förh. Kongl. Svenska Vetensk.-Akad. **1891**(3): 153 (1891). —N.E. Brown in F.T.A. **8**: 19 (1901). —Malme in Bot. Jahrb. Syst. **48**: 304 (1912); in Svensk Bot. Tidskr. **6**: 559 (1912). —Lock in F.T.E.A., Xyridaceae: 21 (1999); in Kew Bull. **54**: 321 (1999). —Lisowski *et al.* in F.A.C., Xyridaceae: 97, fig.42 (2001). Type: Nigeria, Nupe, 1857–1859, *Barter* 764 ?S holotype, K, B, M).

Xyris multicaulis N.E. Br. in F.T.A. **8**: 20 (1901). Type: Malawi, Namasi, *Cameron* 51 (K holotype).

Annual herb, forming small tufts. Leaves linear, 4–8(15) × 0.1–0.15 cm, glabrous, sometimes rugulose towards base, gradually tapering to weakly curved apex. Peduncle up to 35 cm long, 0.5 mm in diameter, weakly 2-ridged particularly towards apex, sometimes weakly rugulose on ridges. Inflorescence narrowly ovoid to narrowly ellipsoid, 4.5–6.5 mm long. Sterile bracts 4, ovate, 3–4 × 1.7–2.5 mm, acute, apiculate, pale brown with darker centre and apex, keeled towards apex; fertile bracts similar but larger, c.5 × 3 mm. Lateral sepals oblong, strongly keeled, c.5 × 1.2 mm, pale brown, keel entire, brown, darker particularly towards apex; third sepal membranous, forming a hood over bud. Corolla yellow, tube c.2 mm long; lobes oblong, c.2 × 1 mm. Anthers 0.5 mm long; filaments ligulate, 0.5 mm long; staminodes much-branched. Style-branches c.1 mm long; stigmas much-branched and lobed. Seeds broadly ellipsoid, c.0.35 × 0.2 mm, with 14–20 longitudinally ridges.

Zambia. B: Zambezi (Balovale), 8.vii.1963, *Robinson* 5579 (K, SRGH). N: Kasama Dist., 80 km N of Kasama, 29.iv.1962, *Robinson* 5133 (K, SRGH). W: Kitwe Dist., Mufulira Nature Reserve, 3.v.1934, *Eyles* 8212 (K, SRGH). S: Choma Dist., 35 km N of Choma, 17.v.1954, *Robinson* 759 (K). **Zimbabwe**. N: Makonde Dist., near Banket, 22.ii.1994, *Seine* 964 (K). W: Matobo Dist., Farm Besna Kobila, viii.1955, *Miller* 2969 (K, SRGH). C: Harare Dist., Hunyani Poort, 2.v.1952, *Wild* 3813 (K, LISC, SRGH). E: Mutare Dist., Rowa, Zimunya's Reserve, 16.vii.1961, *Chase* 7508 (K, SRGH). **Malawi**. N: Rumphi Dist., Nyika Plateau, Chowo Rocks, 17.v.1970, *Brummitt* 10870 (K, SRGH). S: Mulanje Dist., 16 km NW of Likabula Forestry Depot, 15.vi.1962, *Robinson* 5356 (K).

Widespread in the more seasonal parts of tropical Africa from Nigeria to South Africa. Shallow soil over rock, flushes, seasonal swamps among grass tussocks; 700–1400 m.

Conservation notes: Widespread in sites with shallow soils such as rock outcrops; not threatened.

This is the most easily distinguished annual species. The narrowly ovoid or ellipsoid inflorescences with their ovate, acute pale brown bracts, distinguish it from other annual taxa. The leaf sheaths are usually somewhat rugulose, at least towards the base, and the scape is usually weakly 2-ridged towards the apex. The plants are generally small but basal branching can lead to the formation of substantial tufts. The types of *X. straminea* and *X. multicaulis* are similar and I here regard them as the same taxon. Unfortunately the name *X. multicaulis* has been much applied to a form of *X. capensis*.

31. **Xyris rubella** Malme in Bot. Jahrb. Syst. **48**: 303 (1912). —Hepper in Kew Bull. **21**: 424 (1968). —Lock in F.T.E.A., Xyridaceae: 21 (1999); in Kew Bull. **54**: 321 (1999). Types: Namibia, Okahandja, 11.iv.1909, *Dinter* 944 (B, S, SAM syntypes); Tanzania, Kilwa Dist., Orero to Kilwa Karingi, 4.vi.1906, *Braun* 1326 (EA, S syntypes).

Xyris capensis 'paedogenic variant' sensu Lewis & Obermeyer in F.S.A. **4**(2): 2 (1985).

Small annual herb, forming loose tufts. Leaves in two rows, 2–6(14) × 0.1–0.15 cm, glabrous, rugulose, apex solid, apiculate to aristate. Peduncle 8–20(30) cm long, 0.5 mm in diameter, rugulose, 2-winged, particularly towards apex. Inflorescence subglobose, c.3 mm in diameter. Lowest bracts broadly elliptic, 1.3 × 1 mm; sterile bracts broadly spathulate, c.1.8 × 1.3 mm, apiculate, smooth, wine-red to reddish-brown; fertile bracts broadly elliptic, c.2.7 × 2.4 mm, keeled, pale red-brown with darker keel, smooth, apiculate. Lateral sepals arcuate, folded, pale brown, 2.2 × 0.5 mm, apiculate, keel smooth. Corolla yellow, tube 2 mm long, lobes suborbicular, c.1.8 mm wide. Stamens 0.8 mm long; filaments 0.4 mm long; anthers 0.4 mm, divergent at base; staminodes bifurcate, c.0.5 mm long. Ovary obovoid, 1.5 mm long; style 1 mm long, branches 1 mm long; stigma branches recurved. Seeds ellipsoid, 0.3–0.15 mm, with 12–14 longitudinal ridges.

Botswana. SE: N of Kgale Siding, 2.iv.1978, *Hansen* 3397 (K, SRGH). **Zambia**. N: Mbala Dist., Abercorn Pans, 26.iii.1957, *Richards* 8872 (K). W: Mufumbwe Dist., 7 km

E of Mufumbwe (Chizera), 27.iii.1961, *Drummond & Rutherford-Smith* 7449 (K, LISC, SRGH). C: Kabwe Dist., Muka Mwanji Hills, 25 km SSW of Kabwe, 25.ii.1973, *Kornaś* 3290 (K). **Zimbabwe**. N: Gokwe Dist., Sengwa Research Station, 16.iv.1969, *Jacobsen* 620 (K). W: Matobo Dist., Matobo Nat. Park, 4.iii.1994, *Seine* 1115 (K). C: Bindura Dist., Chinamora communal land, Ngomakurira, 25.iii.1952, *Wild* 3783 (K, SRGH). E: Mutare Dist., Nyamkwarara (iNyumquarara) Valley, ii.1935, *Gilliland* K1534 (collection mixed with *X. schliebenii*) (K). S: Chivi Dist., near Madzivire Dip, 6 km N of Runde R., 3.v.1962, *Drummond* 7915 (K, LISC, SRGH). **Malawi**. N: Karonga Dist., Kayelekera, 32 km WSW of Karonga, 9.vi.1989, *Brummitt* 18464 (K). C: Lilongwe Dist., 38 km S of Lilongwe on M1, 6.iv.1978, *Pawek* 14325b (K). **Mozambique**. N: Malema Dist., W side of Serra Inago, 20.iii.1964, *Torre & Paiva* 11278 (LISC, LMA, LMU). Z: Ile Dist., near Errego, 2.iv.1943, *Torre* 5048 (LISC).

Widespread in the more seasonal parts of tropical Africa. Shallow soils on rocky outcrops, between grass tussocks in seasonally wet grassland, often with *Xyris straminea* and *Eriocaulon*; many collections are mixed with *X. schliebeni*; 300–1700 m.

Conservation notes: Widespread but in a restricted habitat; not threatened.

This taxon, together with the next, was regarded by Lewis (1981) and Lewis & Obermeyer (1985) as a juvenile variant of *Xyris capensis*. However, its rugulose leaf sheaths, leaf margins and scapes, as well as the aristate leaf apex, distinguish it from both *X. capensis* and *X. schliebenii*. It is almost always very small, few plants exceeding 20 cm in height, and the inflorescence bracts are normally tinged wine-red.

Xyris scabridula Rendle (*Welwitsch* 2470 (BM, K)) found in Angola, Congo and Burundi is very similar, but there are a few blunt cilia on the keel of the lateral sepals, the leaves are much more finely rugulose and the leaf apex is acute, solid and hooked, rather than aristate as in *X. rubella*.

32. **Xyris schliebenii** Poelln. in Ber. Deutsch. Bot. Ges. **61**: 204 (1944). —Lock in F.T.E.A., Xyridaceae: 22 (1999); in Kew Bull. **54**: 322 (1999). Type: Tanzania, Kilwa, Mbuera, 18.vi.1932, *Schlieben* 2435 (B holotype, BM, BR).

Xyris capensis 'paedogenic variant' sensu Lewis & Obermeyer in F.S.A. 4(2): 2 (1985.

Small annual herb to 30 cm, very similar to *Xyris rubella* but differing in the smooth, not rugulose, sheaths and laminas, in the incurved, obtuse, sometimes mucronate (not straight and aristate) lamina and peduncle sheath apices, and in the more strongly keeled emarginate and apiculate fertile bracts.

Zambia. N: Mpika Dist., 80 km S of Kasama, 29.iv.1962, *Robinson* 5132 (EA, K, P, SRGH). W: Kasempa Dist., Kasempa, 7 km E of Chizera, 27.iii.1961, *Drummond & Rutherford-Smith* 7448 (K, SRGH). **Zimbabwe**. W: Matobo Dist., Besna Kobila Farm, iii.1959, *Miller* 5881 (K, SRGH). C: Marondera, 13.iv.1948, *Corby* 75 (K, SRGH). E: Mutare Dist., Nyamkwarara (iNyumquuarara) valley, ii.1935, *Gilliland* K1534 (collection mixed with *X. straminea*) (K). S: Beitbridge Dist., Tshiturupadzi (Chiturupadzi) store, 12.v.1958, *Drummond* 5779 (K, LISC, SRGH). **Malawi**. N: Rumphi Dist., Nyika Plateau, Chowo Rock, 15.iv.1975, *Pawek* 9237 (K). **Mozambique**. N: Meconta Dist., between Nampula and Corrane, 11.iv.1937, *Torre* 1390 (LISC).

Also in Tanzania and Angola, probably widespread in the more seasonal parts of tropical Africa but its distribution is unclear due to confusion with the previous species; many collections are mixed. Seasonally wet grassland, roadsides, margins of seasonal pools and marshes, often on or around rock outcrops; 300–2200 m.

Conservation notes: Widespread species; not threatened.

This species is extremely similar to *Xyris rubella* but differs from that species in the smooth, not rugulose, sheaths and laminas, in the incurved, obtuse, sometimes mucronate (not straight and aristate) lamina and peduncle sheath apices, and in the more strongly keeled emarginate and apiculate fertile bracts. It appears to be more plastic than *X. rubella* and specimens from Tanzania attain a height of 40 cm. *Drummond & Rutherford-Smith* 7448 (*X. schliebenii*) and 7449 (*X. rubella*) were collected at the same time in the same place and noted as differing in bract colour, showing the need for very careful collecting of these small species.

33. **Xyris parvula** Malme in Bot. Jahrb. Syst. **48**: 304 (1912). —Lock in F.T.E.A., Xyridaceae: 22 (1999); in Kew Bull. **54**: 322 (1999). Type: Tanzania, Mafia Is., Tirene, 10.iii.1909, *Kränzlin* 2983 (B holotype, EA).

Small annual herb. Leaves 2–3 × 0.1 cm; sheaths smooth, glabrous, brown, gradually dilated to the base, rather abruptly tapering above; ligule absent or up to 0.5 mm long; lamina linear, smooth, glabrous; apex acute to obtuse, incurved. Peduncle 6–13 cm long, smooth, glabrous, terete or slightly 2-ridged towards apex. Mature inflorescence ellipsoid, 3–4 × 1.5–2 mm. Outer sterile bracts very broadly ovate, concave, 1.5 × 1.3 mm; inner sterile bracts suborbicular, concave, 2 × 2 mm; fertile bracts suborbicular, concave, c.2.5 × 2.5 mm, brown. Lateral sepals arcuate, 2.5 × 0.5 mm, acute, keel entire. Anthers c.1 mm long, sagittate at base; filaments c.1 mm long. Ovary c.1.2 mm long; style 1.5 mm long, 3-branched. Seeds ellipsoid (only immature ones seen), c.0.25 mm long, with 12–14 longitudinal ridges.

Mozambique. N: Angoche Dist., Angoche (Antonio Enes), near Matangula beach, 20.x.1965, *Mogg* 32424 (K, LISC, SRGH). MS: Muanza Dist., Cheringoma Coast, Nyamaruza dambo, between Camp and Chiniziua lighthouse, v.1973, *Tinley* 2902 (LISC, SRGH).

Also along the East African coast in Kenya and Tanzania. Seasonally wet habitats near the coast; 0–50 m.

Conservation notes: The coastal habitat makes it vulnerable to development for tourism; probably Lower Risk near threatened.

Xyris parvula is very close to *X. schliebenii* but differs in its inflorescence, which is dark brown and subspherical. The leaf sheaths are usually slightly rugulose. Further investigation could well show that this is no more than a coastal ecotype of *X. schliebenii*, in which case the name *X. parvula* would have priority.

34. **Xyris fugaciflora** Rendle in Cat. Afr. Pl. Welw. **2**(1): 71 (1899). —N.E. Brown in F.T.A. **8**: 18 (1901). —Lock in Kew Bull. **54**: 322 (1999). —Lisowski *et al.* in F.A.C., Xyridaceae: 89, fig.37 (2001). Type: Angola, Pungo Andongo, between Caghuy and Sansamanda, 1.v.1857, *Welwitsch* 2461 (BM lectotype, selected by Lock 1999, K).

Small annual herb. Stems solitary, 15–25(30) cm tall. Leaves flattened, 4–7 × c.0.1 cm; lamina weakly rugulose particularly towards the base, margins papillose; apex acute, asymmetric. Peduncle 15–24 cm tall, 0.1–0.2 cm in diameter, terete, slightly flattened above, with 2 marked ridges and other lesser ones. Inflorescence obovoid, truncate; outermost bracts broadly elliptic, 2 × 1.5 mm, straw-yellow; fertile bracts broadly obovate, 3 × 2 mm, straw-yellow. Lateral sepals 3.5 × 0.8 mm, narrowly obovate, asymmetrical; keel entire, glabrous. Flowers not seen. Ovary c.1.5 mm long. Capsule ellipsoid, 3 × 1.5 mm. Seeds ellipsoid, 0.3 × 0.2 mm, with 10–12 longitudinal ridges and similar cross ridges, sometimes irregularly reticulate.

Zambia. W: Mwinilunga, 7 km N of Kalene Hill, 16.iv.1965, *Robinson* 6581 (K).

Also in Angola and Congo. Damp soil over rock outcrops; c.1400 m.

Conservation notes: Known from only a few widely scattered collections; Data Deficient.

The specimens originally cited by Rendle (*Welwitsch* 2461, 2462, 2464) are not uniform. Lewis informally selected *Welwitsch* 2461 as the lectotype and this is confirmed here. The combination of weakly to moderately rugulose leaves and broadly ellipsoid to subspherical inflorescences with straw-yellow bracts distinguishes it from other annual taxa.

Lisowski (2001) took a broad view of this taxon and cites specimens that I would place under *X. rubella* and *X. schliebenii.*

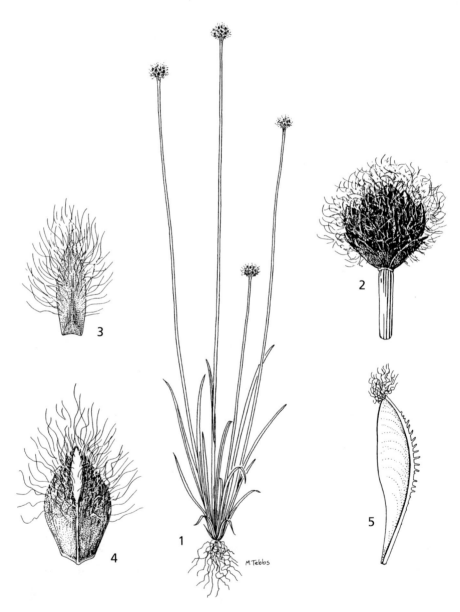

Fig. 13.4.3. XYRIS LANICEPS. 1, habit (× ²/₃); 2, inflorescence (× 4); 3, outer (sterile) inflorescence bract (× 9); 4, inner (fertile) inflorescence bract (× 9); 5, lateral sepal (× 9). All from *Richards* 9306. Drawn by Margaret Tebbs. From Kew Bulletin.

35. **Xyris laniceps** Lock in Kew Bull: **54**: 323 (1999). Type: Zambia, Kawambwa Dist.,
    Ntambachusi (Timnatushi) Falls, 18.iv.1957, *Richards* 9308 (K holotype, BR).
    FIGURE 13.4.**3**.

Small annual herb. Leaves up to 7 × 0.2 cm. Sheaths c.2.5 cm long, smooth or weakly ridged,
green with hyaline margins; ligule short, rounded, hyaline. Lamina up to 4.5 cm long, flat,
smooth, asymmetric and aristate at apex. Scapes up to 21 cm tall, 0.5 mm in diameter, terete,
with weak longitudinal ridges. Inflorescence subspherical to broadly ellipsoid, c.6 × 6 mm when
mature. Lowest pair of sterile bracts 4.2 × 1.4 mm, narrowly obovate, pale brown, darker in
middle, with subapical elongated greyish dorsal mark, margins arachnose-ciliate; second pair 4
× 2.5 mm, broadly ovate, with faint dorsal mark; margins arachnose-ciliate; fertile bracts broadly
ovate, 5 × 3 mm, brown with paler margins, concave, with subapical narrowly ovate dark grey-
green dorsal area, minutely papillose particularly above, arachnose-ciliate towards apex. Lateral
sepals c.4 × 1 mm, arcuate, keeled, pale brown, darker on keel, keel spinose-dentate,
particularly towards the middle. Flower, fruit and seeds not known.

Zambia. N: Kawambwa Dist., Ntambachusi (Timnatushi) Falls, imm.fl. 18.iv.1957,
Richards 9306 (K).
Known only from N Zambia. Top of waterfalls in very wet white sand; 1260 m.
Conservation notes: Known only from the type collection; possibly Endangered.
The only collection is immature, but this species is extremely distinctive in the
whitish arachnoid pubescence of the inflorescences. Further collections would be
most welcome.

ERIOCAULACEAE

by S.M. Phillips

Annual or perennial herbs usually under 1 m high, often much smaller; stem usually
abbreviated to a basal disc, infrequently elongate. Leaves narrow, lanceolate to filiform,
spirally arranged, crowded into a basal rosette or rarely dispersed on elongate stems, opaque
to translucent, sometimes septate. Inflorescence capitate, single or in umbels, on leafless
ribbed scapes. Capitula composed of many small (often tiny) unisexual flowers on a central
receptacle surrounded by one to several whorls of involucral bracts, monoecious, each flower
often subtended by a floral bract. Flowers trimerous or less often dimerous,
subactinomorphic to strongly zygomorphic. Perianth usually composed of two distinct
whorls, sometimes reduced or absent; calyx of free, partially or almost completely joined
sepals, sometimes spathe-like, especially in the male flowers; petals free or partially joined,
spathulate to filiform, frequently hairy, often with a subapical black gland; male petals often
fused with the floral axis to form a funnel-shaped structure topped by very small free lobes.
Male flowers with stamens as many or twice as many as the sepals; a vestigial glandular
gynoecium usually present. Female flowers with a superior 2–3-locular ovary, style simple,
tipped with 2, 3 or 6 elongate stigmas. Fruit a thin-walled, loculicidally dehiscent capsule,
each locule containing a single seed.

The family comprises about 10 genera found throughout the tropics and subtropics
in marshy or seasonally inundated places, especially in montane areas on sand; rarely
truly aquatic. The greatest concentration of genera and species is found in upland
parts of South America; *Eriocaulon* is the only genus in Asia.
The family is easily recognizable by its capitula of small crowded flowers, on top of
leafless scapes and surrounded by involucral bracts. The capitula are reminiscent of
Compositae but, unlike that family, are never brightly coloured, occurring in shades
of white, dirty grey, brown or black.

1. Stamens 4 or 6, twice as many as petals; petals usually with a black gland near or at the tip· 2
– Stamens 3, as many as petals; petals usually eglandular · · · · · · · · · · · · · · · · 3
2. Petals of female flowers free; leaves, scapes and sheaths usually glabrous · · · · ·
· **1. Eriocaulon**
– Petals of female flowers joined into a tube, only bases free; leaves, scapes and sheaths usually hairy· **2. Mesanthemum**
3. Petals of female flowers free · 4
– Petals of female flowers joined in the middle, bases and tips free; leaves, scapes and sheaths usually hairy and often glandular· · · · · · · · · · · · · **3. Syngonanthus**
4. Leaves, scapes and sheaths glabrous; petals of male flowers small but distinct· ·
· **1. Eriocaulon**
– Leaves, scapes and sheaths hairy; petals of male flowers absent · · · · · · · · · · ·
· **4. Paepalanthus**

1. ERIOCAULON L.

Eriocaulon L., Sp. Pl.: 87 (1753). —Ruhland in Engler, Pflanzenr. **13**: 30–117 (1903). —Hess in Ber. Schweiz. Bot. Ges. **65**: 115–204 (1955). —Phillips in Kirkia **17**: 12–49 (1998).

Annuals or perennials, leaves and scapes usually glabrous. Leaves crowded in a basal rosette, or dispersed along elongate stems in floating aquatic species, lanceolate to filiform, septate (divided internally by transverse septa). Scapes unbranched, often twisted, enclosed at the base by a tubular sheath. Involucral bracts of capitulum thin and dry to tough, spreading or reflexed at maturity; floral bracts frequently white-hairy towards tip; flowers sessile or pedicellate on the central receptacle, male and female mixed or the female round the periphery (especially in annuals). Sepals free or ± connate, especially in male flowers, usually free in the female, often boat-shaped, sometimes winged. Petals with a subapical or apical black gland on the inner face, or the glands sometimes reduced or absent, often white-hairy at tip, male petals all very small or one enlarged and exserted from the capitulum. Stamens twice as many as petals (except *E. angustibracteum*); filaments of inner whorl joined to petals; anthers black or white. Seeds relatively large, ellipsoid, yellow to reddish-brown, smooth or distinctively patterned.

About 400 species found worldwide in tropical and subtropical regions; 1 species in temperate E USA and W Europe. The exact number of species is very uncertain, but speciation is well developed throughout the whole range.

Identification of most *Eriocaulon* species is not easy due to the very small size of the flowers, which nevertheless exhibit a wide range of important differences. Frequently species of extremely similar external appearance have widely different floral morphology; only with a few of the most distinctive species is it safe to name a specimen without first dissecting the capitulum and examining the flowers. This particularly applies to the small ephemerals, often with black glabrous capitula, which spring up round the margins of drying pools and runoffs. Several species often grow together in these locations, and care must be taken to avoid mixed gatherings.

The flowers easily hydrate in a drop of water for dissection without the need for boiling. Measurements of floral parts are taken from hydrated flowers. Other measurements refer to dried material; leaves and scapes often shrink considerably on drying. It is important where possible to examine mature capitula as the dimensions of the floral parts can alter drastically at maturity. In strongly zygomorphic flowers the longest petals elongate rapidly at anthesis in the male flowers and when seed is ripe in the female, and wings on the sepals increase in size. Leaves usually taper to some extent from the sheathing base to the tip and width measurements are taken about halfway up. Various types of hairs occur on the floral parts, and short, thick, obtuse, very opaque white hairs are widespread. Parts bearing such hairs are described here as white-papillose when the hairs are short, or white-pilose or white-villous when they are longer.

The patterning of the seed coat is of great importance in *Eriocaulon*, both for identification and for establishing relationships. Sometimes species with almost identical floral morphology have very different seeds. Seed-set can be sparse in the robust perennials, but otherwise seed is usually abundantly produced and is available for inspection. Although a high magnification is needed to see fine detail, the coarse patterning can be discerned with a hand lens, which is often sufficient to confirm an identification.

Key to species

1. Submerged aquatics, only the capitula emergent; plant with a long leafy stem or well-developed rhizome; leaves numerous, floating · · · · · · · · · · · · · · · · · · 2
 – Marsh plants or ruderals; plant tufted, stem short or absent; leaves basal (rarely aquatic but without a long stem or rhizome) · 5
2. Anthers black; leaves numerous, usually filiform, 0.1–0.5(4) mm wide · · · · · · 3
 – Anthers white; leaves linear, 2–8 mm wide · 4
3. Capitula blackish, 3–5 mm wide; viviparous capitula and stolons absent; female petals ± glabrous · **1.** *setaceum*
 – Capitula dirty white, 5–8 mm wide; viviparous capitula and stolons usually present; female petals white-villous · **2.** *ramocaulon*
4. Viviparous capitula present, arching down and rooting as stolons; leaves in a rosette, rhizome absent; capitula 4–9 mm wide · · · · · · · · · · · · · · **3.** *africanum*
 – Viviparous capitula absent; leaves clustered at apex of stout rhizome; capitula 8–13 mm wide · **4.** *latifolium*
5. Anthers white · 6
 – Anthers black · 8
6. Perennial, often robust; capitula 4–9 mm wide · · · · · · · · · · · · · · **3.** *africanum*
 – Slender rosulate annuals; capitula 2–4 mm wide · 7
7. Female flowers with 0–3 filiform caducous sepals; petals absent · · · · **5.** *cinereum*
 – Female flowers with 2 oblong sepals; petals well developed · · · · · · · · **6.** *varium*
8. Flowers with 2 petals, 4 anthers and 2 stigmas · 9
 – Flowers with 3 petals, (3)6 anthers and 3 stigmas · · · · · · · · · · · · · · · · · · · 10
9. Seeds uniformly brown; sepals of female flowers narrowly to very broadly winged · **7.** *mutatum*
 – Seeds brown with transverse white stripes; sepals of female flowers with a wing not wider than the sepal-body · **8.** *maronderanum*
10. Flowers with 2 sepals and 3 petals in both male and female flowers; sepals narrow, unwinged · 11
 – Flowers with 3 sepals and 3 petals · 13
11. Seeds transversely white-papillose; petals of female flowers subglabrous, eglandular · **9.** *fuscum*
 – Seeds coarsely reticulate with longitudinal bands; petals of female flowers pilose or marginally ciliate, glandular · 12
12. Petals of female flowers pilose on the face, eglandular or gland small; capitula shiny white to yellowish-grey, c.5 mm wide; floral bracts obovate · · · · · · **10.** *truncatum*
 – Petals of female flowers ciliate on margins, tipped with a large black gland; capitula dark grey-brown, 3.5–4 mm wide; floral bracts oblong · · · **11.** *schlechteri*

13. Petals ciliate on margins with long translucent hairs, tipped with a large black gland; seeds longitudinally ridged · **11.** *schlechteri*
 – Petal margins not ciliate with long translucent hairs; seeds not longitudinally ridged· 14
14. Perennials from a short rhizome, usually robust; capitula 8–20 mm wide, densely white-hairy · 15
 – Small ephemerals, annuals, occasionally perennials (but then capitula < 7 mm wide or leaves < 8 cm long); capitula 1.5–9 mm wide, glabrous or hairy · · · · 20
15. Sepals of female flowers ciliate along inside margins; leaves thin, finely acute · **12.** *sonderianum*
 – Sepals of female flowers not ciliate along inside margins (bearded at base in *E. mesanthemoides*); leaves usually thick, spongy· 16
16. Leaves not usually exceeding 20 cm long, often much shorter, 1–5 mm wide; scapes 4–8-ribbed; capitula 8–12 mm wide · 17
 – Leaves large, thick, to 35 cm long and 4–18 mm wide; scapes stout, 8–10-ribbed; capitula 9–20 mm wide · 19
17. Sepals of female flowers fused into a spathe · · · · · · · · · · · · · · · · · · · **13.** *pictum*
 – Sepals of female flowers free · 18
18. Marsh plant; leaves up to 30 cm long and 4 mm wide, spongy, tipped with a pore · **14.** *teuszii*
 – Aquatic, leaves up to 50 cm long, c.1 mm wide, translucent; forming floating mats · **15.** *taeniophyllum*
19. Female sepals glabrous inside, sometimes fused and spathe-like; receptacle glabrous; involucral bracts slightly shorter than capitulum width, coriaceous at least towards base· **16.** *schimperi*
 – Female sepals bearded inside; receptacle pilose; involucral bracts as wide as capitulum, scarious · **17.** *mesanthemoides*
20. Sepals of female flowers subequal, similar · 21
 – Sepals of female flowers very unequal, of different form, the two laterals with a distinct spongy or winged keel, median sepal much narrower · · · · · · · · · · · 35
21. Involucral bracts thin, soon crumpling, not visible in mature capitulum; scape 4-winged; capitula black with a covering of short white hairs · · · · **18.** *elegantulum*
 – Involucral bracts clearly visible in mature capitulum; scape-ribs rounded · · · 22
22. Sepals and petals of female flowers linear to filiform; male sepals free; capitula 5–8 mm wide, shaggy, pale and shining · **19.** *bongense*
 – Sepals of female flowers narrowly lanceolate or oblong to ovate; male sepals usually joined, at least near base; capitula not as above · · · · · · · · · · · · · · · 23
23. Male calyx acutely 3-lobed, usually from about middle; seeds uniformly brown; small ephemerals to 12 cm high · 24
 – Male calyx lobes obtuse to truncate-denticulate, often almost completely joined; seeds with white reticulate or striate patterning · 26
24. Sepals of female flowers not winged· 25
 – Sepals of female flowers spongily winged · **22.** *wildii*
25. Capitula dark grey to blackish; floral bracts narrowly obovate, acute · **20.** *abyssinicum*
 – Capitula whitish to pale grey; floral bracts narrowly lanceolate, acuminate-aristate· **21.** *welwitschii*
26. Involucral bracts radiating beyond head of flowers · · · · · · · · · · · **23.** *infaustum*
 – Involucral bracts not radiating beyond head of flowers· · · · · · · · · · · · · · · · · 27
27. Male sepals free to the base; leaves tipped with a pore · · · · · · · · · **24.** *matopense*
 – Male sepals at least partially fused; leaves without a pore · · · · · · · · · · · · · · · 28

28. Sepals of female flowers with spongy wing or thickening on the keel, usually ciliate on the margins or along inside edges with translucent hairs; perennial from a short rhizome · 29
– Sepals of female flowers without a conspicuous spongily thickened keel, margins glabrous; annual (except 31. *E. mbalensis*) · 31
29. Sepals of female flowers boat-shaped, margins ciliate, glabrous inside; capitula 5–10 mm wide, white· **12.** *sonderianum*
– Sepals of female flowers gibbous, margins glabrous, usually hairy inside; capitula 4–7 mm wide, greyish-white· 30
30. Capitula subglobose, slightly wider than long, 5.5–7 mm wide, sometimes viviparous; floral bracts oblanceolate-oblong, acute, the inner densely white-hairy; female sepals gibbous with a broad spongy wing wider than sepal body · **25.** *zambesiense*
– Capitula globose to dome-shaped, often longer than wide, 4–5 mm wide, never viviparous; floral bracts broadly spathulate-cuspidate, sparsely white-hairy: female sepals narrowly winged or merely thickened, wing narrower than sepal body · **26.** *inyangense*
31. Involucral bracts scarious, greyish, reflexing at maturity· · · · · · · · · · · · · · · 32
– Involucral bracts firm, straw-coloured, not reflexing at maturity· · · · · · · · · · 33
32. Seeds transversely striate; sepals of female flowers winged, acute · **27.** *transvaalicum*
– Seeds reticulate; sepals of female flowers usually unwinged, truncate-denticulate · **28.** *selousii*
33. Scapes pilose; involucral bracts narrowly lanceolate, acute · · · · · · **31.** *mbalensis*
– Scapes glabrous; involucral bracts oblong to obovate, rounded· · · · · · · · · · · 34
34. Capitula shortly white-hairy; leaves linear, acute; seeds black, white-reticulate · **29.** *afzelianum*
– Capitula glabrous; leaves subulate to filiform, bristle-tipped; seeds brown, white-striate · **30.** *stenophyllum*
35. Anthers 3; capitula 4–6 mm wide; leaves bristle-tipped· · · · · **32.** *angustibracteum*
– Anthers 6; capitula 3–5 mm; leaves not bristle-tipped · · · · · · · · · · · · · · · · 36
36. Leaves subulate, acuminate, up to 12 cm long, much longer than scape-sheaths · **33.** *deightonii*
– Leaves linear, subacute, up to 3 cm long, equalling scape-sheaths· · · · · · · · · 37
37. Floral bracts acute to acuminate; capitula 4–5 mm wide · · · · · · · · · · · · · · · 38
– Floral bracts rounded to subacute; capitula 2.5–4 mm wide · · · · · · · · · · · · · 39
38. Male calyx spathe-like; capitula dark grey; scape slender; seeds with transverse rows of white papillae · **34.** *buchananii*
– Male calyx entire, funnel-shaped; capitula pale grey-brown; scape stiff, fairly stout; seeds white-reticulate · **35.** *chloanthe*
39. Capitula black; median sepal of female flowers well developed; seeds white papillose · **36.** *mulanjeanum*
– Capitula brown-grey; median sepal of female flowers much reduced; seeds white reticulate· **37.** *maculatum*

1. **Eriocaulon setaceum** L., Sp. Pl.: 87 (1753). —Ruhland in Engler, Pflanzenr. **13**: 89 (1903). —Obermeyer in F.S.A. **4**(2): 10 (1985). —Phillips in F.T.E.A., Eriocaulaceae: 5 (1997); in Kirkia **17**: 16 (1998). Type: Sri Lanka, *Herb. Hermann* 1: 40 no.50 (BM lectotype).

   Eriocaulon melanocephalum Kunth, Enum. Pl. **3**: 549 (1841). —Kimpouni, Lejoly & Lisowski in Fragm. Fl. Geobot. **37**: 130 (1992). Type: Brazil, *Sellon* s.n. (K).

Eriocaulon bifistulosum Van Heurck & Müll. Arg. in Van Heurck, Obs. Bot.: 105 (1870). —
N.E. Brown in F.T.A. **8**: 239 (1901). —Ruhland in Engler, Pflanzenr. **13**: 90 (1903). —Hess in
Ber. Schweiz. Bot. Ges. **65**: 130 (1955). Type: Nigeria, Nupe, Jehan, *Barter* 1021 (G, K).

Aquatic herb; stem floating below the surface, slender, branching, up to 50 cm or more
long, densely clothed in numerous leaves, terminating in a stiff umbel of 1–20 scapes
emerging above the water surface. Leaves filiform, yellow-green, flaccid, 3–10 cm long, 0.1–0.3
mm wide, 1-nerved, septate, not sheathing at base. Scapes up to 30 cm high, 5–7-ribbed;
sheaths shorter than leaves, loose, shortly 1–3-fid at mouth. Capitulum subglobose, (2)3–5(6)
mm wide, blackish or white-pubescent; involucral bracts shorter than the capitulum width,
blackish, scarious, obovate-oblong to suborbicular, broadly rounded, weakly reflexing at
maturity; floral bracts cuneate-oblong, blackish, concave, acute, the inner sometimes white-
papillose on the back; receptacle glabrous or thinly pilose; flowers trimerous, 0.9–1.4 mm
long, pedicellate, black. Male flowers: sepals oblong, concave, unequally joined into a spathe
or almost free, glabrous or white papillose near the rounded tips; petals small, glandular;
anthers 6, black. Female flowers: sepals subequal or the median smaller, deeply concave, the
back gibbous with broadly rounded tip, spongy along the midline, infrequently lightly keeled
and winged with a few white hairs on the upper keel and an acute tip, upper margins smooth
or denticulate; petals subequal, linear-spathulate, rather spongy, glands small or absent,
glabrous or a few white papillae at the tip; ovary black, only very tardily dehiscent. Seeds
0.4–0.5 mm long, transversely white papillose-striate, mid-brown but often appearing blackish
due to the adhering ovary wall.

Botswana. N: Okavango, E side of Xere Is., 23.vi.1973, *P.A. Smith* 631 (K, LISC,
SRGH). **Zambia**. N: Kasama Dist., Mungwi, 20.vi.1960, *Robinson* 3759 (K, SRGH). W:
Kalulushi Dist., Luano Forest Res., 23.iii.1966, *Fanshawe* 9655 (K, SRGH). **Zimbabwe**.
N: Guruve Dist., Nyamunyechi Estate, Mabubu dam, 8.vi.1978, *Nyariri* 217 (SRGH).
W: Matobo Dist., Chesterfield Farm, iv.1959, *Miller* 5877 (K, SRGH). C: Harare Dist.,
Twentydales, 28.iv.1948, *Wild* 2523 (K, SRGH).

Widespread throughout the tropics, including Australia. Rooting in the mud of
slow-moving or stagnant water up to 50 cm deep, only the capitula emerging above
the surface; 1000–1500 m.

Conservation notes: Widely distributed; not threatened.

E. setaceum is immediately recognisable by its distinctive aquatic habit, the long
floating stems being clothed in numerous filiform leaves and terminated by an
emergent cluster of often blackish capitula. Usually in *Eriocaulon* the mature seed
is readily expelled from the capitulum by capsule dehiscence, but in *E. setaceum*
the whole fruit is shed, sometimes splitting into 3 mericarps which fall separately.
The seed thus appears blackish due to the adhering ovary wall. The seed,
surrounded by a layer of mucilage, is later expelled from this wall through the
usual loculicidal slit.

E. setaceum sensu lato is one of the most widespread species of *Eriocaulon*, and
comprises a complex of forms. Variation in the structure of the bracts and flowers has
led to the description of several species throughout its range. These differences relate
mainly to the colour of the capitulum, hairiness of the receptacle (glabrous or pilose),
presence of white papillae on the bracts and sepals (when dense the capitulum
appears white), whether the female sepals are obtuse or acute or thickened along the
midline into a wing, and whether the petals are glandular or pilose. The status of these
variants can only be determined by a study covering the whole range of the complex,
so the name is applied here in a broad sense to cover all forms in the Flora area.

E. submersum Rendle differs from *E. setaceum* by its larger capitula (8 mm wide) with
bigger flowers (c.2 mm), and by its broader, often 3-nerved leaves (up to 0.8 mm
wide). It is known only from the type collected in Huíla, Angola and its status is very
doubtful. It may be simply an unusually vigorous collection of *E. setaceum*.

2. **Eriocaulon ramocaulon** Kimpouni in Fragm. Flor. Geobot. **39**: 344 (1994). — Phillips in Kirkia **17**: 17 (1998). Type: Congo, Kundelungu Plateau, 7 km SW of Poste Katohufa, 1630 m, 7.xi.1968, *Malaisse* 6005 (BR holotype).

> *Eriocaulon malaissei* Moldenke forma *viviparum* Moldenke in Phytologia **19**: 345 (1969). Type as above.

Submerged aquatic perennial from an elongate rhizome clothed in roots and old leaves. Leaves numerous, linear-subulate to filiform, 10–15 cm long, up to 4 mm wide but often mostly <0.5 mm, closely nerved, septate, acute, woolly in the axils, matted hairs remaining around the rhizome. Scapes 3–6, stout, 25–32 cm high, 11–12-ribbed; sheaths equalling the leaves, loose, rather thick, the limb splitting into several lobes. Capitulum subglobose, 5–8 mm wide, dirty white, sometimes viviparous; involucral bracts slightly shorter than capitulum width, light brown flushed grey, coriaceous with thinner margins and tip, broadly obovate-oblong, rounded across the tip, glabrous, reflexed at maturity, in 2 or 3 series grading into the floral bracts; floral bracts narrowly oblanceolate-oblong, translucent with a tougher base and midline, pallid flushed grey above, lightly concave with incurving acute tip, white-pilose upwards; receptacle pilose to glabrescent. Flowers trimerous, 2.5–3 mm long. Male flowers: sepals free, pale with grey tips, linear-oblong, the two lateral lightly keeled, median sepal flat, white-pilose towards the obtuse tips; petals shortly exserted from calyx, subequal, glandular and densely white-villous on the inner face; anthers black. Female flowers: sepals as in the male flowers; petals and ovary sessile; petals oblanceolate, subequal, glandular, white-villous on the inner face. Seeds plumply elliptic, 0.5–0.6 mm long, brown with white projections.

Zambia. N: Chinsali Dist., Shiwa Ngandu, 19.vii.1938, *Greenway* 5413 (K); Mbala Dist., Mululwe R., W of Mbala (Abercorn), 22.vii.1933, *Michelmore* 501 (K). W: Mwinilunga Dist., Zambezi R. at rapids c.6.5 km N of Kalene Hill mission, 16.vi.1963, *Edwards* 800 (SRGH).

Also in Congo (Katanga). Growing in rocky beds of swift-flowing upland streams, the leaves submerged and forming dense mats, the capitula emergent above the water surface; 1600–1700 m.

Conservation notes: Moderately widely distributed; probably not threatened.

E. ramocaulon is easily distinguished from *E. africanum*, a similar caulescent aquatic species of upland streams, by its finer leaves and black anthers.

3. **Eriocaulon africanum** Hochst. in Flora **28**: 340 (1845). —N.E. Brown in Fl. Cap. **7**: 56 (1897). —Hess in Ber. Schweiz. Bot. Ges. **65**: 266 (1955). —Obermeyer in F.S.A. **4**(2): 19 (1985). —Kimpouni in Fragm. Flor. Geobot. **39**: 324 (1994). — Phillips in Kirkia **17**: 18 (1998). Type: South Africa, KwaZulu-Natal, Mgeni R. near Pietermaritzburg, ix.1840, *Krauss* 375 (K, BM, MO).

> *Eriocaulon woodii* N.E. Br. in Fl. Cap. **7**: 57 (1897). Type: South Africa, KwaZulu-Natal, near Murchinson, x.1884, *Wood* 3053 (K holotype, NH).
>
> *Eriocaulon antunesii* Engl. & Ruhland in Engler, Bot. Jahrb. **27**: 76 (1899). Type: Angola, Huíla, *Antunes* 139 (B holotype).
>
> *Eriocaulon stoloniferum* Rendle in Cat. Afr. Pl. Welw. **2**: 101 (1899). Type: Angola, Huíla, Morro de Lopollo, v.1860, *Welwitsch* 2458 (BM holotype).
>
> *Eriocaulon woodii* var. *minor* Ruhl. in Engler, Pflanzenr. **4**: 70 (1903). Type: South Africa, KwaZulu-Natal, Umlaas, 1885, *Wood* 524 (B holotype).
>
> *Eriocaulon natalensis* Schinz in Mém. Herb. Boiss. **10**: 76 (1900). Type as for *E. woodii*.
>
> *Eriocaulon malaissei* Moldenke in Phytologia **19**: 343 (1970). —Phillips in Kirkia **17**: 19 (1998). Type: Congo, Katanga, Luanza R., 5 km from source, 1580 m, 3.viii.1966, *Malaisse* 4489 (BR).
>
> *Eriocaulon parvistoloniferum* Kimpouni in Fragm. Flor. Geobot. **39**: 342 (1994). Type: Congo, Katanga, Upemba Nat. Park, R. Muye, *de Witte* 4269 (BR holotype).

Tufted perennial, rarely with a slender stem up to 5 cm long. Leaves linear, septate, very variable in length, 6–35 cm long, 2–5 mm wide, falcately curved when short, longer leaves ribbon-like, flaccid, tapering to a subacute tip. Flowering scape single, 15–40 cm high, straight, stout, 6–10-ribbed, not twisted, usually one or more scapes bearing viviparous capitula also present, these arching downwards and rooting like stolons; sheath loose, shortly slit, the limb broadly rounded, splitting into 2–4 obtuse lobes. Capitulum subglobose, 4–9 mm wide, the white-hairy petals exceeding the blackish floral bracts; involucral bracts shorter than the capitulum width, suborbicular, yellowish, papery, reflexed at maturity; floral bracts oblanceolate-oblong, blackish or brown, scarious, glabrous or the inner with a few white papillae, subacute; receptacle glabrous. Flowers trimerous, 1.7–2.5 mm long. Male flowers: calyx black, variable, sepals almost free, or frequently two joined for most of their length and the third free, all concave, glabrous or white-pilose across the rounded tips and along the midline upwards; petals subequal, the longest up to 1.5 mm long, all glandular and densely white-pilose; anthers creamy-white. Female flowers: sepals black, deeply concave but not keeled, almost glabrous to thinly villous with white hairs along the centre back, or sometimes lower hairs translucent and extending on to the pedicel, the margins often irregularly or coarsely toothed; petals linear-spathulate, the median slightly larger, glandular and white-pilose at tips, villous with white or translucent hairs on the inner blade. Seeds plumply ellipsoid, 0.6–0.7 mm long, very pale brown and translucent, almost smooth with a faint reticulate patterning.

Zambia. N: Mbala Dist., Fwambo area, Fisa R. near gorge, 3.ix.1956, *Richards* 6078 (K, SRGH). C: Serenje Dist., Kundalila Falls c.13 km SE of Kanona, 15.x.1967, *Simon & Williamson* 1001 (K, LISC, SRGH). **Zimbabwe**. C: Marondera Dist., Macheke, 11.ix.1940, *Hopkins* in SRGH 7727 (K). E: Chimanimani Dist., Chimanimani Mts, foot of Mt Peza, 12.x.1950, *Wild* 3581 (K, LISC); Nyanga Dist., Nyanga Nat. Park, Mare R., below Rhodes Inyanga Expt. Station, ix.1975, *Burrows* 834 (K, SRGH). **Mozambique**. MS: Sussundenga Dist., Makurupini R., 23.ix.1986, *Müller & Bluemel* 3827 (SRGH).

Also in South Africa (KwaZulu-Natal, Limpopo, Mpumalanga), Angola and Congo (Katanga). Submerged or at the edge of swiftly flowing upland streams and rivers, rooting in sand among rocks on the river bed, sometimes in rock pools by waterfalls. When submerged it may form grassy looking patches or mats, with only the flowering heads emergent above the water. Found in moving water rather than marshy ground; 900–1800 m.

Conservation notes: Widely distributed; not threatened.

The species is easily recognized by its pale anthers and viviparous capitula acting as stolons. It is extremely variable in vigour, probably depending on environmental conditions.

4. **Eriocaulon latifolium** Sm. in Rees, Cyclop.: 13 (1819). —N.E. Brown in F.T.A. **8**: 243 (1901). —Ruhland in Engler, Pflanzenr. 4(30): 78 (1903). —Meikle in F.W.T.A. ed.2, **3**: 62 (1968). Type: Sierra Leone, *Afzelius* s.n. (LINN holotype, BM, UPS).

Eriocaulon rivulare Benth. in Hook., Fl. Niger: 547 (1849). Type: Sierra Leone, *Vogel* s.n. (K holotype).

Eriocaulon thunbergii Körn. in Linnaea **27**: 677 (1856). Type: Sierra Leone, *Afzelius* s.n. (S holotype, B).

Eriocaulon banani Lecomte in Bull. Soc. Bot. Fr. **55**: 645 (1908). Type: Mali, Banan, 4.iii.1899, *Chevalier* 524 (P holotype, BR, K).

Eriocaulon vittifolium Lecomte in Bull. Soc. Bot. Fr. **55**: 645 (1908). Types from Guinea and Mali.

Robust aquatic perennial, rosette-forming when young, stem gradually elongating to form a tough rhizome densely clothed in roots and fibrous leaf remnants. Leaves clustered at rhizome apex, linear, often ribbon-like and trailing, up to 40 cm long, 3–8 mm wide, many-veined, bases expanded, pale, papery, tips subacute. Scapes 1–6, erect, stout, 7–10-ribbed, not twisted, up to 35 cm high; sheaths shorter than the leaves, very loose, mouth obliquely slit, limb acuminate, soon lacerate. Capitulum hemispherical becoming subglobose, 8–13 mm wide, colour variable, the white-hairy petals intermingled with neatly imbricate pallid or dark floral bracts; involucral bracts shorter than the capitulum width, suborbicular to broadly obovate-oblong, yellowish, firmly papery, merging into the floral bracts; outer floral bracts oblanceolate-oblong, rounded, grading into narrowly oblong acute inner bracts, incurving, pallid, bronze or dark grey, usually glabrous or the inner white-pilose on the back; receptacle flattened, glabrous. Flowers trimerous, 2.2–3.5 mm long, pedicellate. Male flowers: sepals variably joined into a spathe or infrequently almost free, pallid to dark grey, oblanceolate, concave, white-pilose across the rounded tips and along the midline upwards, sometimes sparsely; petals subequal, glandular, densely white-pilose, longest 0.5–1.7 mm; anthers creamy-white. Female flowers: sepals pallid to dark grey, obovate to oblanceolate-oblong, all concave or the median slightly narrower and almost flat, not keeled, upper back sparsely white-pilose, tips rounded, white-pilose, margins and tip entire or crenulate-denticulate; petals and ovary raised on a stipe, petals narrowly oblanceolate, clawed, the median slightly larger, glandular or sometimes glands inconspicuous or absent on one or more petals, densely white-villous on upper part of inner blade, densely to sparsely villous with translucent hairs below. Seeds plumply ellipsoid, pale brown, faintly reticulate, 0.4–0.6 mm long.

Zambia. N: Kasama Dist., c.10 km N of Kasama, 3.vii.1930, *Hutchinson & Gillett* 4033 (K, SRGH); Mporokoso, 12.ix.1958, *Fanshawe* 4806 (K, LISC).

Also in Angola, Congo (Katanga), Cameroon and W Africa. Submerged in fast-flowing upland streams and rooting among rocks on the stream bed, leaves trailing in the water current and the capitula emergent above the water surface; 1250–1500 m.

Conservation notes: Fairly widely distributed; probably not threatened.

E. latifolium is variable in small details of floral morphology, including colour of the floral bracts, relative length of sepals and petals, shape of the sepals and hairiness of the petals in the female flowers, and in the presence or absence of petal glands. Several species names have been based on these differences, but they are not correlated with distribution and appear to be simply minor variations.

E. latifolium is very similar to *E. africanum* in floral and seed morphology, but is generally a much bigger, more vigorous plant with broader leaves and larger capitula. It lacks the striking stolon-like arching scapes with rooting viviparous capitula of *E. africanum*, but instead spreads by elongation of the stem, which in time forms a tough rhizome with roots along its length. Typical specimens of these two species are easily distinguished, but where their distribution overlaps in Zambia the distinction is less clear. A few rather vigorous specimens with arching 'stolons' also possess an elongate, root covered rhizome, e.g. *Symoens* 12609 (Zambia C: Serenje).

5. **Eriocaulon cinereum** R. Br., Prodr. Fl. Nov. Holl.: 254 (1810). —Obermeyer in F.S.A. **4**(2): 11 (1985). —Phillips in F.T.E.A., Eriocaulaceae: 6 (1997); in Kirkia **17**: 19 (1998). Type: Australia, *R. Brown* s.n. (BM holotype).

Eriocaulon amboense Schinz in Bull. Herb. Boiss. **4**, App. 3: 35 (1896). —N.E. Brown in F.T.A. **8**: 258 (1901). —Ruhland in Engler, Pflanzenr. **13**: 112 (1903). —Hess in Ber. Schweiz. Bot. Ges. **65**: 176 (1955). —Friedrich-Holzhammer & Roessler in Merxmüller, Prodr. Fl. SW Afr., fam.159: 1 (1967). Type: Namibia, Ovambo, Uashitenga near Olukonda, viii.1885, *Schinz* 859 (Z holotype, K).

Eriocaulon stuhlmanni N.E. Br. in F.T.A. **8**: 259 (1901). Type: Tanzania, Mwanza Dist., E Uzinza, *Stuhlmann* 3552 (B holotype).

Small tufted annual. Leaves narrowly linear to needle-shaped, bristle-tipped, forming a neat dense cluster, 1–3 cm long, 0.2–2 mm wide, septate. Scapes slender, often numerous, 4–13 cm high, 5-ribbed; sheaths ± equalling the leaves, subinflated, mouth scarious, shortly obliquely slit. Capitulum globose to ovoid, 2–4 mm wide, grey- and straw-coloured or becoming darker, the bracts very loosely erect; involucral bracts as wide as the capitulum, obovate-oblong, rounded becoming ragged on margins, ascending, pallid, scarious becoming tougher downwards; floral bracts lanceolate-oblong, thinly scarious, pale with a dark central band or sometimes blackish, glabrous or a few inner lightly keeled with fine sparse hairs on the keel; receptacle thinly pilose. Flowers trimerous, pedicellate. Male flowers: sepals joined into a spathe with tridentate tip, dark grey, glabrous; petals included within the calyx, tiny; anthers white, 0.1 mm long. Female flowers: much reduced and consisting mostly of the gynoecium; sepals 3, 2 or absent, filiform, caducous; petals absent, their position indicated by a node on the ovary stipe. Seeds 0.3 mm long, light brown, glossy, faintly reticulate.

Botswana. N: Chobe Dist., Tsotsoroga, 18.v.1977, *P.A. Smith* 2022 (K, SRGH). **Zambia**. B: Sesheke, 18.vi.1963, *Fanshawe* 7865 (SRGH). W: Kasempa Dist., 7 km E of Chizera, 27.iii.1961, *Drummond & Rutherford-Smith* 7690 (SRGH). **Zimbabwe**. N: Gokwe Dist., Sengwa Research Station, 6.iv.1968, *Jacobsen* 164 (SRGH). W: Matobo Dist., Besna Kobila, iv.1957, *Miller* 4342 (K, SRGH).

In scattered localities across West Africa from Senegal to Chad and in Tanzania, Angola and N Namibia; probably not native in Africa. India to China, Japan, SE Asia and Australia; introduced to warm areas elsewhere as a weed of rice cultivation. Drying pool margins and seepages in lowland areas; 1000–1400 m.

Conservation notes: Widely distributed; not threatened.

A small species, distinctive on account of its tiny white anthers and reduced female flowers; the cushion of small needle-shaped leaves is also characteristic.

6. **Eriocaulon varium** Kimpouni in Fragm. Flor. Geobot. **39**: 357 (1994). —Phillips in Kirkia **17**: 20 (1998). Type: Congo, Katanga, Upemba Nat. Park, near Lusinga, *Lisowski, Malaisse & Symoens* 4750 (POZG holotype, BR, BRVU).

Tufted annual. Leaves subulate, 2–5 cm long, 0.5–1.5 mm wide, thin, septate, tapering to a bristle tip. Scapes up to 10, slender, slightly flexuous, 14–20 cm high, 6-ribbed; sheaths equalling the leaves, obliquely slit. Capitulum subglobose with a flattened base, 3–4 mm wide, white and grey, bracts loose, the flowers visible between; involucral bracts obovate-oblong, broadly obtuse, firm, yellowish with scarious grey margins and tip, not reflexing at maturity; floral bracts narrowly oblong, obtuse to subacute, translucent, concave and flushed grey above the middle, glabrous or thinly white-pilose on upper back, lightly keeled below and ciliate with short colourless hairs on the lower keel; receptacle hemispherical, glabrous or pilose. Flowers trimerous, 1.3–1.7 mm long. Male flowers: sepals joined into a dark grey spathe, the lateral lightly keeled, free tips white-pilose; petals on a slender, parallel-sided stipe, small, subequal, glandular and white-pilose; anthers whitish. Female flowers: sepals 2, grey, narrowly oblong to boat-shaped, lightly keeled, tips acute and glabrous or white-pilose, shortly ciliate with colourless hairs on the keel; petals and ovary sessile or raised on a stipe; petals 3, subequal, narrowly oblong-spathulate, clawed, tips glandular and white-pilose, some colourless hairs on the blade. Seeds ellipsoid, brown, 0.6 mm long, faintly reticulate.

Zambia. N: Ndola, 2.iii.1963, *Fanshawe* 7718 (K, SRGH). W: Mwinilunga Dist., 7 km N of Kalene Hill, 16.iv.1965, *Robinson* 6667 (K).

Also in S Congo (Katanga) and Tanzania. In shallow water of pans and pools on rock outcrops; 1400–1600 m.

Conservation notes: Moderately widely distributed; probably not threatened.

E. varium is best distinguished from other slender erect annuals by its pale anthers and seeds lacking white papillae. It is most closely related to *E. cinereum*, which is usually smaller and lacks petals in the female flowers.

7. **Eriocaulon mutatum** N.E. Br. in F.T.A. **8**: 256 (1901), based on *E. huillense* Rendle (May 1899). —Hess in Ber. Schweiz. Bot. Ges. **65**: 167 (1955). —Phillips in F.T.E.A., Eriocaulaceae: 9 (1997); in Kirkia **17**: 21 (1998). Type: Angola, Huíla, between Lopollo and Monino, *Welwitsch* 2448 (BM syntype, K), 2449 & 2450 (BM syntypes).

Small rosulate annual. Leaves subulate to filiform, 1–3.5 cm long, 0.2–1 mm wide, tapering to a slenderly acuminate or bristle tip. Scapes 10–50, filiform, 3–15(22) cm high, 2–4-ribbed; sheaths shorter or longer than the leaves, inflated upwards and deeply slit, the limb obtuse to acute, sometimes splitting. Capitulum globose to slightly dome-shaped, 1.5–3(4.5) mm wide, black, the bracts loose with flowers visible among them; involucral bracts about as wide as the capitulum, grey or infrequently light brown (lighter than rest of capitulum), scarious, obovate to subrotund with rounded tip, spreading at maturity; floral bracts obovate to obovate-oblong, blackish, glabrous, obtuse to acute; receptacle glabrous or thinly hairy. Flowers dimerous, blackish, 0.7–1 mm long, completely glabrous. Male flowers: sepals free, narrowly oblanceolate-oblong, obliquely ragged-truncate; petals unequal, very small and included within the calyx, eglandular, one 0.3 mm long, emarginate, the other vestigial; anthers black. Female flowers: sepals very variable, concave and keeled, broad with gibbous overlapping margins and winged keel, the wing scarious, up to 0.6 mm wide with a coarsely toothed margin and the upper sepal margins also toothed, varying to narrowly falcate and unwinged, or at most with a small tooth at the centre back, tip acute to cuspidate; petals dimorphic, eglandular, the dorsal oblanceolate, 1 mm long, entire to emarginate, the ventral shorter and narrower, linear-oblong, bidentate. Seeds ellipsoid, 0.3 mm long, brown, smooth.

This delicate annual, although seldom collected, occurs over a wide area of southern tropical Africa. The slender habit with small black capitula is very similar to that of *E. abyssinicum*, but it can be immediately distinguished from this by its dimerous flowers. Specimens with narrow unwinged female sepals look strikingly different from the more usual form with a broad toothed wing, but every gradation exists between the two extremes. A distinction at specific level is untenable, but forms with ± unwinged sepals often have somewhat smaller capitula and filiform leaves and are distinguished here at varietal level. Additionally, a form from Zambia with a more robust habit than usual, larger capitula and extremely broadly winged sepals, is also distinguished as a variety.

1. Female sepals broad and enclosing the seed, keel winged, wing to 0.8 mm wide, coarsely toothed; capitulum 2.5–4.5 mm wide; leaves usually subulate, 0.4–1.2 mm wide · 2
– Female sepals narrow and exposing the seed, keel unwinged or at most with a tooth < 0.1 mm wide; capitulum 1.5–2.5(3) mm wide; leaves often filiform, 0.2–0.3 mm wide · var. *angustisepalum*
2. Capitula 2.5–3.5 mm wide; scapes to 15 cm high; wing on female sepals 0.2–0.5 mm wide, flowers suborbicular · var. *mutatum*
– Capitula 3.5–4.5 mm wide; scapes 10–22 cm high; wing on female sepals 0.5–0.8 mm wide, flowers wider than long · var. *majus*

Var. **mutatum** —Phillips in Kew Bull. **51**: 643 (1996); in F.T.E.A., Eriocaulaceae: 9 (1997). FIGURE 13.4.**4**.

Eriocaulon huillense Rendle in Cat. Afr. Pl. Welw. **2**: 95 (May 1899) non Engl. & Ruhland (Apr. 1899). Type as for *E. mutatum*.

Eriocaulon pseudomutatum Kimpouni in Fragm. Flor. Geobot. **39**: 343 (1994). Type: Congo, Katanga, near source of Luansoko, *Duvigneaud* 2954e (BRLU holotype).

Zambia. B: Zambezi Dist., c.21 km W of Zambezi (Balovale), 26.v.1960, *Angus* 2276 in part (mixed with var. *angustisepalum*) (SRGH). N: Kawambwa Dist., Kawambwa, 21.vi.1957, *Robinson* 2332 (K, SRGH). W: Kasempa Dist., Chizera, 100 km WNW of Kasempa, 7.vii.1963, *Robinson* 5553 (K, SRGH). **Zimbabwe**. C: Marondera Dist., Marondera (Marandellas), 13.iv.1948, *Corby* 80 (K, SRGH). **Malawi**. N: Chitipa Dist., Nyika Plateau, 14.5 km down Nthalira road, 5.vii.1973, *Pawek* 6998 (K). C: Nkhotakota (Kota Kota), 2.v.1963, *Verboom* 105 (K). S: Zomba Mt, near Ngongola Forestry Village, 8.iv.1971, *Jordan* 3002 (K).

Also in S Tanzania, S Congo and Angola. Dambos, wet grassland, ditches and in seepage areas over rocks; 1000–2000 m.

Conservation notes: Widely distributed; not threatened.

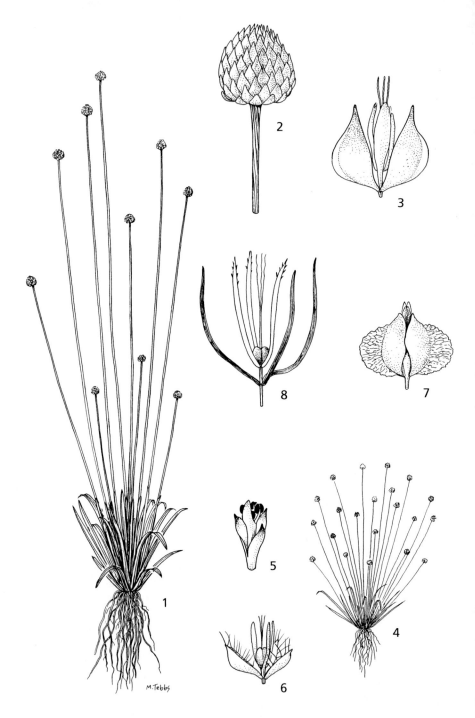

Fig. 13.4.4. ERIOCAULON BUCHANANII. 1, habit (× ²/₃); 2, capitulum (× 4); 3, female flower (× 14). E. ABYSSINICUM. 4, habit (× ²/₃); 5, male flower (× 14); 6, female flower (× 14). E. MUTATUM var. MUTATUM. 7, female flower (× 14). E. BONGENSE. 8, female flower (× 14), 1–3 from *Faden et al. 96/482 & 96/97*, 4–6 from *Faden et al. 96/396*, 7 from *Milne-Redhead & Taylor 9890*, 8 from *Faden et al. 96/329*. Drawn by Margaret Tebbs. From Flora of Tropical East Africa.

Var. **angustisepalum** (H. Hess) S.M. Phillips in Kew Bull. **51**: 644 (1996). —Phillips in F.T.E.A., Eriocaulaceae: 10 (1997). Type: Angola, Cuanhama, road verges 45 km S of Cassinga, *Hess* 52/2004 (ZT holotype, BR, K).

> *Eriocaulon angustisepalum* H. Hess in Ber. Schweiz. Bot. Ges. **65**: 170 (1955). — Obermeyer in F.S.A. **4**(2): 11 (1985).

Zambia. B: Zambezi Dist., c.21 km W of Zambezi (Balovale), 26.v.1960, *Angus* 2276 in part (mixed with var. *mutatum*) (SRGH). N: Kasama Dist., 96 km E of Kasama, Kalungu R. terrace, 19.v.1964, *Astle* 3044 (SRGH). C: Serenje Dist., 1 km from Kanona on road to Kundalila Falls, 12.iii.1975, *Hooper & Townsend* 675 (K). **Zimbabwe**. C: Harare (Salisbury), Hatfield, 8.v.1934, *Gilliland* 86 (K, SRGH).

Also in S Tanzania, Angola, South Africa (Gauteng, Limpopo, KwaZulu-Natal). A pioneer taxon in sandy, marshy places, seepage areas along road verges and over rocks; 1000–1700 m.

Conservation notes: Widely distributed; not threatened.

Var. **majus** S.M. Phillips in Kew Bull. **51**: 644 (1996). Type: Zambia, Northern Prov., Chishinga Ranch near Luwingu, 15.v.1961, *Astle* 632 (K holotype).

Zambia. N: Luwingu Dist., Chishinga Ranch near Luwingu, 15.v.1961, *Astle* 632 (K). W: Mwinilunga Dist., Solwezi road, 76 km from Mwinilunga, 14.v.1972, *Kornaś* 1783 (K). C: Mkushi Dist., 53 km NE of Kabwe (Broken Hill), 15.vii.1930, *Hutchinson & Gillett* 3647 (K).

Only known from these areas. Shallow margins and muddy edges of drying pools, marshy ground, ditches; 1250–1500 m.

Conservation notes: Localised distribution; probably Near Threatened.

8. **Eriocaulon maronderanum** S.M. Phillips in Kew Bull. **51**: 646 (1996); in Kirkia **17**: 23 (1998). Type: Zimbabwe, Marondera Dist., in vlei, 15.vii.1948, *Corby* 134 (K holotype, SRGH).

Annual, the slender scapes rising above the small basal leaf rosette. Leaves linear to lanceolate, 1–2 cm long, 0.8–2 mm wide, spongy, acute. Scapes up to 50, 6–11 cm high, 3–4-ribbed; sheaths equalling the leaves, the limb entire or becoming toothed. Capitulum globose to ovoid, 3–4 mm wide, black, slightly glossy, the flowers and bracts loosely intermingling; involucral bracts oblong, rounded, grey, firmly scarious, as wide as the capitulum in a single series, reflexing at maturity; floral bracts obovate or obovate-oblong, black, acute; receptacle pilose. Flowers dimerous, 1.2–1.7 mm long, black, glabrous or a very few white papillae on the sepal and petal tips. Male flowers: sepals free, oblanceolate, lightly keeled, obliquely truncate-denticulate; petals very small, eglandular; anthers (3)4, black. Female flowers: sepals oblong-falcate, winged on the keel, the wing about as wide as sepal body, wing margin sometimes toothed, tip acute; petals unequal, the larger narrowly oblanceolate-oblong, obtuse, the smaller emarginate; ovary stipitate. Seeds ellipsoid, 0.4 mm long, brown with white-papillose reticulations or striae.

Zimbabwe. C: Harare (Salisbury), 2.v.1931, *Brain* 3736 (K, LISC, SRGH). **Malawi**. N: Mzimba Dist., Katoto, 5 km W of Mzuzu, 3.viii.1973, *Pawek* 7278 (K, MO, SRGH).

Only known from the Flora area. In wet upland dambo grassland; 1400–1500 m.

Conservation notes: Apparently rather localised distribution; probably Near Threatened.

E. maronderanum is very similar morphologically to *E. mutatum*, but with an altogether different seed testa patterning. *E. mutatum* has a smooth seed and usually a wider wing on the female sepals and the two are probably not closely related.

9. **Eriocaulon fuscum** S.M. Phillips in Kew Bull. **51**: 629 (1996); in Kirkia **17**: 24 (1998). Type: Zimbabwe, Hwange Dist., Matetsi Safari Area, 22.v.1978, *Gonde* 210 (K holotype, SRGH). FIGURE 13.4.**7**.

Delicate rosulate annual. Leaves subulate, spongy, 1–1.2 cm long, 1–1.5 mm wide, acute. Scapes c.20, subfiliform, much exceeding the small leaf rosette, 8–10 cm high, 3-ribbed; sheaths equalling the leaves, inflated upwards, obliquely slit with acuminate limb. Capitulum subglobose, 3–3.5 mm wide, greyish-brown, bracts loosely erect; involucral bracts as wide as the capitulum, light brown, scarious, narrowly obovate to oblong, obtuse with denticulate upper margin, not reflexed; floral bracts resembling those of the involucre in colour and texture, oblong, cuspidate, the innermost tipped with a few white papillae; receptacle glabrous. Flowers 1 mm long. Male flowers: sepals 2, free, grey, narrowly oblong, falcate, acute; petals rudimentary; stamens 6 but usually only 4 or 5 anthers developed, black. Female flowers: sepals 2(3), grey, linear, falcate, glabrous or a few small white papillae on upper margins, caducous; petals 3, raised on a short stipe, linear, eglandular, a few white papillae on upper margins, median petal a little larger, its tip denticulate; ovary sessile, trilocular. Seeds ellipsoid, 0.3 mm long, pale yellow and translucent becoming light brown, transversely white-papillose.

Zimbabwe. W: Hwange Dist., Matetsi Safari Area game fence, LK 845724, 22.v.1978, *Gonde* 210 (K, SRGH).

Known only from the type. Wet hollow in a dambo on light grey clay soil; c.950 m.

Conservation notes: Very localised distribution but within a protected area; probably Vulnerable.

In floral and seed morphology *E. fuscum* most closely resembles *E. polhillii* S.M. Phillips from Tanzania, but the latter lacks petals in the female flowers. *E. fuscum* also has a very much smaller basal rosette of leaves and smaller capitula with less obvious involucral bracts. To the casual eye it is easily confused with other small dark-headed annual species from Zimbabwe, especially *E. cinereum*, *E. mutatum* and *E. abyssinicum*, but these are all very different in flower details. *E. fuscum* is most easily distinguished from *E. cinereum* by its black anthers, from *E. mutatum* by its trilocular ovary, and from *E. abyssinicum* by its free male sepals and linear female sepals.

10. **Eriocaulon truncatum** Mart. in Wallich, Pl. As. Rar. **3**: 29 (1832). —Phillips in F.T.E.A., Eriocaulaceae: 11 (1997); in Kirkia **17**: 24 (1998). Type: India, Bihar, Monghir (Munger) Hills, *Hamilton* s.n., Wallich cat.no. 6076 (K-WA holotype).

    Eriocaulon annuum Milne-Redh. in Hooker's Icon. Pl. **34**: t.3389 (1939). Type: Tanzania, Mafia Is., Dawe Simba to Ndaagoni, 4.x.1937, *Greenway* 5389 (K holotype).

    Eriocaulon ciliipetalum H. Hess in Ber. Schweiz. Bot. Ges. **65**: 263 (1955). Type: Tanzania, Mafia Is., Ngombeni, 12.vii.1932, *Schlieben* s.n. (Z holotype).

Tufted annual. Leaves linear to linear-lanceolate, up to 6 cm long, 1.5–3 mm wide, opaque, subacute and apiculate. Scapes many, stiff, stout, up to 15 cm high, 5-ribbed; sheaths equalling the leaves, inflated, obliquely slit, the limb acuminate, often splitting and becoming bifid. Capitulum crateriform, c.5 mm wide, glistening white to yellowish-grey; involucral bracts as wide as the capitulum, oblong with a rounded or obtusely triangular tip, shiny, scarious, pallid, spreading at maturity; floral bracts obovate, rounded, similar to those of the involucre but usually shorter; receptacle subglabrous to pilose. Flowers with trimerous ovary, 1.5–1.8 mm long, sepals usually blackish. Male flowers: sepals 2, joined into a 2-lobed spathe, rarely almost free, oblong, lightly keeled, truncate-erose, glabrous or a few white papillae near tips; petals 3, tiny, glandular and white-papillose; anthers black. Female flowers: sepals 2, linear or narrowly oblong, geniculate, folded, glabrous; petals and ovary raised on a stipe up to 0.4 mm long; petals 3, narrowly oblanceolate, subequal, thinly pilose to subglabrous (densely pilose in Africa) on the inner face with septate, translucent hairs, glandular and white-papillose at the tip. Seeds ellipsoid, 0.5 mm long, pale yellow to brownish-yellow, coarsely reticulate with longitudinal bands.

Mozambique. MS: Muanza Dist., Cheringoma coast, Nyamaruza dambo, halfway between camp and road junction to Chiniziua lighthouse, v.1973, *Tinley* 2904 (LISC, SRGH).

Introduced. Widespread in tropical Asia at lower altitudes, often as a weed of rice. In acid bogs; near sea level.

Conservation notes: Widely distributed; not threatened.

The few specimens of *E. truncatum* known from the eastern coast of Africa have more densely hairy female petals than Asian plants, but are otherwise indistinguishable. Seeds with longitudinally elongated cells are extremely unusual in *Eriocaulon*. They are plentifully produced and, coupled with the pale shiny capitula, are the easiest diagnostic character for this species.

11. **Eriocaulon schlechteri** Ruhland in Bot. Jahrb. Syst. **27**: 78 (1899). —N.E. Brown in F.T.A. **8**: 255 (1901). —Phillips in Kirkia **17**: 25 (1998). Type: Mozambique, Inhambane, xi.1897, *Schlechter* 12093 (B holotype, P).

Eriocaulon ruhlandii Schinz in Bull. Herb. Boiss., sér.2 **6**: 710 (1906). —Obermeyer in F.S.A. **4**(2): 14 (1985). Type: South Africa, KwaZulu-Natal, near Clairmont, 18.vii.1893, *Schlechter* 2955 (Z holotype, K, PRE).

Tufted annual. Leaves few, 2–5 cm long, 1–2 mm wide, narrowly linear, thin, acute. Scapes 1–5, 9–16 cm high, slender, 5-ribbed; sheaths subequal to leaves. Capitulum hemispherical, light to dark greyish-brown, 3.5–4 mm wide, glabrous, slightly glossy; involucral bracts narrowly obovate-oblong, as wide as the capitulum, obtuse to subacute, not reflexed; floral bracts resembling the involucral bracts but narrower, glabrous, the upper margins denticulate; receptacle villous with silky, spirally twisted hairs, these untwisting when wet. Flowers with trimerous ovary, 1.1–1.2 mm long. Male flowers: sepals 2 or 3, joined only at base, grey, glabrous, narrowly obovate-oblong, falcate, concave, the median when present straight and flat, tips irregularly denticulate; petals 3, tiny, gland-tipped, the median with a few translucent hairs; anthers 6, black. Female flowers: sepals 2 or 3, free, otherwise resembling the male, median sepal variable, almost equalling the laterals or reduced or absent; petals and ovary sessile; petals 3, subequal, linear, pallid, fleshy, tipped by a large black gland, the median usually with some apical white papillae, margins ciliate above the middle with long, twisted, translucent, septate hairs. Seeds elliptic, yellow or yellowish-brown, 0.35–0.4 mm long, longitudinally ridged.

Mozambique. GI: Inhambane, 2.xi.1897, *Schlechter* 12093 (B holotype).

From S Mozambique southwards to South Africa (KwaZulu-Natal, Eastern Cape). On marshy ground in coastal areas; 5–50 m.

Conservation notes: Only known from one historic specimen in the Flora area, but moderately widely distributed elsewhere; not threatened.

Both male and female flowers may possess 2 or 3 sepals in different flowers from the same capitulum. *E. schlechteri* is most closely related to *E. truncatum*, which shares the unusual seed patterning of longitudinally thickened cells and also has 2–3 sepals. It is easily distinguished, however, by its larger pale capitula. The silky cilia on the petals resemble the receptacular hairs, and untwist when wet.

12. **Eriocaulon sonderianum** Körn. in Linnaea **27**: 669 (1854). —N.E. Brown in Fl. Cap. **7**: 55 (1897). —Ruhland in Engler, Pflanzenr. **13**: 75 (1903). —Phillips in Kirkia **17**: 27 (1998). Types: South Africa, Gauteng, Magaliesberg, *Zeyher* 1731 (W syntype, K); Gauteng, Magaliesberg, *Burke* s.n. (B syntype, K).

Eriocaulon decipiens N.E. Br. in F.T.A. **8**: 245 (1901). Type: Malawi, Mt Mulanje, 1891, *Whyte* 115 (K holotype, BM).

Eriocaulon dregei Hochst. var. *sonderianum* (Körn.) Oberm. in Bothalia **13**: 450 (1981); in F.S.A. **4**(2): 18 (1985).

Rosulate perennial from a branching rootstock, forming colonies. Leaves subulate, 2–7(12) cm long, 1–4 mm wide, slightly spongy, acuminate to a fine point. Scapes 1–3 per rosette, 5–25(35) cm high, 6-ribbed; sheaths about as long as the leaves, the mouth shortly spathate with a rounded limb, or sometimes divided into a circlet of 4 spreading lobes. Capitulum globose or slightly flattened, 5–12 mm wide, dirty white, the base intruded at maturity; involucral bracts shorter than the capitulum width, c.3 mm long, firmly scarious, obovate-oblong, pallid or straw-coloured, the outer obtuse, glabrous, the inner with broadly triangular tips and sparsely white-pilose, reflexed at maturity; floral bracts oblanceolate, scarious, spongy along the keel, flushed grey, densely white-hairy upwards, acute; receptacle hemispherical, villous. Flowers trimerous, 2.5–3.8 mm long. Male flowers: sepals joined below into a funnel-shaped spathe, grey, 2 lobes lightly keeled, the third flat, rounded and white-pilose at the tip; petals shortly exserted from calyx, unequal, white-pilose on the inner face; anthers black. Female flowers: sepals oblong to boat-shaped, deeply concave, grey, scarious, ciliate on the margins with long translucent hairs, keel spongy, sometimes expanded into a wing, white-pilose towards the tip; petals and ovary sessile; petals narrowly oblanceolate-oblong, two 1.7–2.2 mm long, the third slightly larger, white-pilose upwards on the inner face. Seeds 0.6 mm long, broadly ellipsoid, mid-brown.

Zimbabwe. E: Nyanga Dist., Nyangani, Chipunga stream, 26.xi.1949, *Chase* 1874 (BM, K, SRGH). **Malawi**. S: Mulanje Mt, c.2000 m., 11.ii.1958, *Chapman* 424 (BM) & 491 (K). **Mozambique**. MS: Gorongosa Dist., Gorongosa Mt, Gogogo summit area, i.1972, *Tinley* 2330 (LISC, SRGH).

Also in South Africa (Limpopo, Mpumulanga, KwaZulu-Natal, Eastern Cape), Lesotho and Swaziland. Stream margins in montane areas; 1800–2000 m.

Conservation notes: Widely distributed; not threatened.

E. sonderianum is primarily a South African species, reaching its northernmost limit on Mt Mulanje. It is closely related to *E. dregei* Hochst., also from South Africa, of which it is sometimes regarded as a variety. However, on average *E. dregei* is larger with longer leaves (up to 30 cm × 8 mm) and bigger capitula (8–13 mm). The two can always easily be distinguished by their leaf-tips, which are very obtusely rounded in *E. dregei* and finely pointed in *E. sonderianum*.

13. **Eriocaulon pictum** Fritsch in Bull. Herb. Boiss., sér.2 **1**: 1102 (1901). —Phillips in F.T.E.A., Eriocaulaceae: 11 (1997); in Kirkia **17**: 27 (1998). Type: Angola, Huíla, *Dekindt* 703 (W holotype, LISC).

Eriocaulon amphibium Rendle in J. Linn. Soc., Bot. **27**: 475 (1906). Type: Zimbabwe, Matobo Hills, near American Mission, ix.1905, *Gibbs* 210 (BM holotype, K).

Perennial forming clusters of leaf rosettes from a short rhizome clothed in roots and old leaves. Leaves lanceolate, 4–10 cm long, 3–5 mm wide, thick and spongy, yellow-green, tapering uniformly to an acute, hard, yellow-brown tip, woolly in the axils with long matted hairs. Scapes few, often solitary, 25–65 cm high, 6–8-ribbed; sheaths usually clearly longer than the leaf rosette, loose, the mouth scarious, obliquely slit with a rounded limb or becoming ragged. Capitulum subglobose, 8–12 mm wide, grey becoming white; involucral bracts shorter than the capitulum width, c.3 mm long, yellow, brown or coppery, cartilaginous, broadly elliptic to oblong with obtuse or ragged tips; outer floral bracts resembling those of the involucre, narrowly oblong-cuneate, acuminate, grading into grey inner bracts, these scarious with a thicker midline, white-hairy below the cuspidate tip; receptacle pilose, sometimes sparsely. Flowers trimerous. Male flowers: c.5 mm long; sepals narrowly oblong, the two laterals keeled, third flat, joined into a grey spathe with free, white-hairy, subacute tips; petals ligulate, exserted from calyx, unequal with the longest up to 3.5 mm long, white-villous at tips, white-pilose with shorter hairs on the inner blade; anthers black. Female flowers: sepals resembling the male calyx, joined into a grey spathe with free, white-hairy tips; petals raised on a short villous stipe, unequal, narrowly oblanceolate, white-hairy at tips, conspicuously villous with spreading, translucent, septate hairs c.1 mm long from the lower blade, the longest 3 mm long, clawed and clearly exserted from calyx, the other two 2 mm long; ovary sessile. Seeds ellipsoid, 0.75 mm long, brown with short projections.

Zambia. N: Kasama Dist., Mungwi, 8.ii.1960, *Fanshawe* 686 (K). W: Mwinilunga Dist., Dobeka bridge, 11.xii.1937, *Milne-Redhead* 3613 (K, LISC). **Zimbabwe**. W: Matobo Dist., Besna Kobila Farm, i.1956, *Miller* 3322 (K). C: Near Harare (Salisbury), 4.ix.1932, *Eyles* 7177 (K). E: Nyanga Dist., communal land road N of Nyanga village, 28.i.1958, *Norman* R21 (K). **Malawi**. N: Nyika Plateau, Lake Kaulime, 23.x.1958, *Robson & Angus* 293 (K, LISC, SRGH).

Also in Angola and S Tanzania. Seasonal pools and swampy ground, often in peat bogs in association with *E. teucszii*; 1500–2200 m.

Conservation notes: Widely distributed; not threatened.

E. pictum is often confused with *E. teucszii*; the two species are extremely similar in general appearance and also frequently grow together in the same peaty habitats. They can easily be distinguished by the female flowers as the spathate female calyx of *E. pictum* has a characteristic brush of woolly hairs protruding from the slit, and also by the leaf-tip which is obviously rounded in *E. teucszii*, although care is needed as the pointed, brown leaf-tips of *E. pictum* are easily broken off. Leaves of *E. pictum* are yellow-green rather than blue-green as in *E. teucszii*.

A spathate female calyx is very unusual in African *Eriocaulon*, although it occurs occasionally in *E. schimperi*, a much more vigorous species from montane areas above 2000 m, and in *E. lanatum* Hess from Angola, a smaller species lacking a rhizome and with woolly hairs on the scape.

14. **Eriocaulon teusczii** Engl. & Ruhland in Bot. Jahrb. Syst. **27**: 77 (Apr. 1899). — Phillips in F.T.E.A., Eriocaulaceae: 12 (1997); in Kirkia **17**: 28 (1998). Type: Angola, Malange, ix.1879, *Rensch* in *von Mechow* 231 (B holotype, Z).

> *Eriocaulon huillense* Engl. & Ruhland in Bot. Jahrb. Syst. **27**: 78 (Apr. 1899), non Rendle (May 1899). Type: Angola, Huíla, v.1895, *Antunes* s.n. (B holotype).
> *Eriocaulon lacteum* Rendle in Hiern, Cat. Afr. Pl. Welw. **2**: 99 (May 1899). Types: Angola, Huíla, Lopollo, *Welwitsch* 2452; Morro de Lopollo, *Welwitsch* 2452b; Humpata Dist., slopes of Serra de Oiahoia, *Welwitsch* 2453 (all BM syntypes, K, LISU, M).

Perennial from a short rhizome, often robust. Leaves many, rush-like, narrow, semi-cylindrical, up to 20 cm long, 1–3 mm wide, usually crowded in an erect tuft, subulate when short, spongy, the tip obtuse with a pore on the upper surface. Scapes 1–6(15), 25–65 cm high, 6–8-ribbed; sheaths equalling or longer than the leaves, mouth shortly slit, limb rounded or becoming split. Capitulum hemispherical becoming globose with intruded base at maturity, white, 8–12 mm wide, the pointed tips of floral bracts usually visible among the exserted white-hairy petals; involucral bracts shorter than the capitulum width, in several series, spreading, firm, straw-coloured or light brown and often flushed grey or black with a central paler stripe, oblong to ovate, outer rounded, the inner acute; floral bracts cuneate, white-villous on the widest part, firm and straw-coloured below, the tip papery, dark grey, brown or infrequently pale, often abruptly long-cuspidate but sometimes merely acute or apiculate; receptacle pilose. Flowers trimerous, 3–4.5 mm long. Male flowers: sepals free, ± equal, laterals narrowly oblong to boat-shaped, white or with greyish or brownish tips, white-villous towards the obliquely truncate tip, median similar but flat; petals unequal, glandular, white-villous over the whole inner blade, two small, the third narrowly oblong, up to 3 mm long, exserted from the capitulum; anthers black. Female flowers: sepals as in the male flowers; petals and ovary both stipitate; petals unequal, glandular, white-villous over the inner blade, narrowly oblanceolate-oblong, somewhat narrowed to the the base but not clawed, two 1.5–2.5 mm, the third c.$^1/_3$ longer, exserted from capitulum. Seeds subrotund, 0.5–0.6 mm long, reddish-brown with white papillae.

Zambia. B: Sesheke Dist., Kale, 16.iv.1963, *Symoens* 10289 (K). N: Chinsali Dist., Shiwa Ngandu, 19.vii.1938, *Greenway* 5414 (K). W: Kalulushi Dist., Luano Forest Reserve, 17.iii.1966, *Fanshawe* 9623 (SRGH). C: Serenje Dist., Kundalila Falls, 14.x.1963, *Robinson* 5716 (K, SRGH). **Zimbabwe**. W: Matopos, no locality, v.1954,

Garley 1066 (SRGH). C: Marondera (Marandellas), 23.xi.1945, *Rattray* 1371 (SRGH). E: Mutasa Dist., Mt Nusa (Nuza), 25.vi.1934, *Gilliland* 507 (K, SRGH). **Malawi.** N: Lake Malawi, Nkhata Bay, 6.vii.1976, *Pawek* 11476 (K). **Mozambique.** Without locality, 23.v.1919, *W.P. Johnson* s.n. (K).

Also from S Congo (Katanga), S Tanzania, Angola and N Nigeria. Common and sometimes locally dominant in peat bogs, wet swamps among coarse grass and in ditches, sometimes in shallow standing water; 450–2000 m.

Conservation notes: Widely distributed; not threatened.

E. teuszii is vegetatively a very variable species, especially in leaf length. Populations also differ in the colour of the involucral and floral bracts and sepal tips, inconstant characters on which earlier workers based new species. The floral bracts are thinly to densely white-hairy and usually long-cuspidate. When dark-coloured the long tips are conspicuous among the white flowers in the capitulum. The capitula are often dark at first, but become fluffy-white as the petals are exserted. The combination of large fluffy-white capitula, together with unequal petals, free truncate-tipped male sepals, and obtuse leaves tipped with a pore, distinguishes *E. teuszii* from the other robust species. *E. teuszii* is often confused with *E. pictum*, with which it sometimes grows (see note under *E. pictum*).

15. **Eriocaulon taeniophyllum** S.M. Phillips in Kew Bull. **51**: 632 (1996); in Kirkia **17**: 29 (1998). Type: Zambia, Lake Bangweulu, Chiluwi Is., 8.vii.1972, *Verboom* 2750 (K holotype, SRGH). FIGURE 13.4.5.

Tufted aquatic perennial. Leaves all basal, long and ribbon-like, flaccid, c.45 cm long, 0.8–1 mm wide, translucent, septate. Scapes 70–75 cm high, 6-ribbed, only slightly twisted; sheaths shorter than the leaves. Capitulum hemispherical, 12 mm wide, fluffy-white; involucral bracts shorter than capitulum width, in several series, firm with thinner margins and tip, grey-brown with a paler central stripe, outer broadly ovate-oblong, rounded, the inner oblong, subacute, finally reflexing at maturity; floral bracts narrowly oblong-cuneate, incurved, white-villous on the upper back, acute; receptacle glabrous. Flowers trimerous, 4–5.3 mm long. Male flowers: sepals free, subequal, pale with grey tips, linear-oblanceolate, lightly keeled, the laterals falcate, all white-villous towards the obliquely truncate tip; petals seated on a stipe as long as the calyx, glandular, densely white-villous over the inner blade, two small, the median much larger, c.3 m long, oblong, exserted from capitulum; anthers black. Female flowers: sepals resembling the male, oblong to boat-shaped, lightly keeled, spongy along the midline, the laterals falcate, median straight, white-villous on the upper keel; petals and ovary both shortly stipitate; petals glandular, two narrowly oblanceolate-oblong to a narrowed base, white-villous at the tips, the median much larger, narrowly oblong, 3–4 mm long, white-villous over the inner blade above the middle, exserted from capitulum. Seed ellipsoid, brown with white papillae (only immature seed seen).

Zambia. N: Samfya Dist., Lake Bangweulu, Chilubi Is., 8.vii.1972, *Verboom* 2750 (K, SRGH). C: Serenje Dist., Lake Lusiwasi, 17.ix.1959, *Vesey-FitzGerald* 2642 (SRGH).

Not known elsewhere. In water up to 50 cm deep at the swampy margins of lakes, sometimes abundant and forming colonies; 1200–1400 m.

Conservation notes: Endemic to the Lake Bangweulu basin. Localised distribution; probably Vulnerable.

The species closely resembles *E. teuszii* in floral and seed structure, but differs markedly in its truly aquatic habit, reflected in the long flaccid leaves.

16. **Eriocaulon schimperi** Ruhland in Bot. Jahrb. Syst. **27**: 80 (1899). —Kimpouni in Fragm. Flor. Geobot. **39**: 347 (1994). —Phillips in F.T.E.A., Eriocaulaceae: 15 (1997); in Kirkia **17**: 31 (1998). Type: Ethiopia, Jan Meda (Dschan-Mèda), x.1863, *Schimper* 1217 (B holotype, BM, K, P).

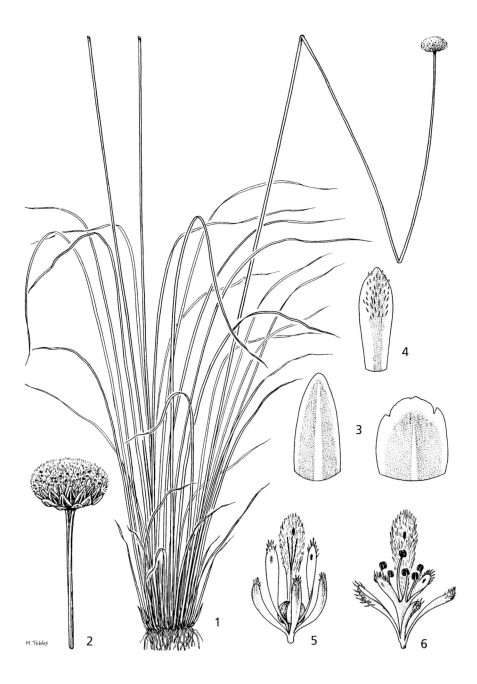

Fig. 13.4.**5.** ERIOCAULON TAENIOPHYLLUM. 1, habit (× ²/₃); 2, capitulum (× 2); 3, involucral bracts (× 10); 4, floral bract (× 10); 5, male flower (× 10); 6, female flower (× 10), all from *Verboom* 2750. Drawn by Margaret Tebbs. From Kew Bulletin.

Eriocaulon volkensii Engl. var. *mildbraedii* Ruhland in Mildbraed, Wiss. Ergebn. Deutsch. Zentr.-Afr. Exp., Bot. **2**: 57 (1910). Type: Congo, *Mildbraed* 1690 (B holotype).

Eriocaulon congense Moldenke in Phytologia **2**: 218 (1947). Type: Congo, Kivu Dist., W base of Mt Mikeno, 2200 m, 20.vi.1927, *Chapin* 404 (NY holotype).

Robust tufted perennial from a basal rootstock or short rhizome. Leaves broadly linear, 6–25(40) cm long, 4–14(18) mm wide, light green, thick, smooth, slightly glossy, the tip broadly rounded. Scapes 2–5, stout, straight, 15–35(60) cm high, 8–10-ribbed; sheaths shorter than the leaves, limb inflated, papery. Capitulum globose or slightly flattened, 9–14(18) mm wide, pure white or greyish-white; involucral bracts slightly shorter than the capitulum width, 2.5–4 mm long, pallid, flushed grey or straw-coloured with blackish tips, firmly scarious or coriaceous with scarious margins and tip, oblong or ovate-oblong, acute, the inner thinly white-pilose, weakly reflexed at maturity; floral bracts oblong-cuneate, grey, densely white-pilose upwards; receptacle glabrous. Flowers trimerous, 3–3.5(4.5) mm long. Male flowers: sepals blackish, connate into a spathe with free tips, obovate-oblong, deeply concave with the two lateral sometimes lightly keeled, broadly acute, white-pilose across the tips or also on the upper back; petals glandular and white-villous, the longest 1.5–2.5 mm long, its tip exserted from the calyx; anthers black. Female flowers: sepals blackish, subequal, oblong to obovate-oblong, often connate at the base, concave, spongy along the midline and occasionally broadened into a small wing above the middle, white-pilose towards the subacute tip; petals and ovary subsessile; petals subequal, ligulate to oblong-spathulate, glandular and white-pilose near the tip, sometimes also translucent hairs below the white ones. Seeds ellipsoid to subrotund, 0.7 mm long, brown.

Malawi. N: Nyika Plateau, Lake Kaulime, 23.x.1958, *Robson & Angus* 273 (K, LISC); Nyika Plateau, Chelinda R. bridge, 10.ix.1976, *Pawek* 11798 (K, SRGH).

Ethiopia, Sudan (Imatong Mts), Congo (eastern mountains), Uganda, Kenya and Tanzania. Wet montane grassland and marshy streamsides; 2200–2500 m.

Conservation notes: Although widely distributed in Africa, apparently restricted to the Nyika Plateau within the Flora area, where it is Near Threatened.

A robust montane species, characterised by a large clump of thick, blunt leaves and a few stout scapes bearing large capitula. It is primarily an East African species, reaching the southernmost point in its range on the Nyika Plateau where it can be abundant, especially around Lake Kaulime. Populations throughout its range show small variations, especially in the texture and colour of the involucral bracts and the degree of hairiness of the floral bracts and flowers leading to differences in capitulum colour. The Nyika plateau population has pure white capitula with pale involucral bracts and densely white-hairy floral bracts.

17. **Eriocaulon mesanthemoides** Ruhland in Bot. Jahrb. Syst. **27**: 79 (1899). —N.E. Brown in F.T.A. **8**: 244 (1901). —Phillips in F.T.E.A., Eriocaulaceae: 15 (1997); in Kirkia **17**: 32 (1998). Type: Tanzania, Morogoro Dist., Uluguru Mts, Ukami, 6.xi.1894, *Stuhlmann* 9143 (B holotype).

Robust tufted perennial from a short rhizome. Leaves narrowly lanceolate, 7–30 cm long, 6–10 mm wide, firm, gradually tapering to a narrowly obtuse tip. Scapes 1–5, stout, 10–60 cm high, 8–9-ribbed; sheaths usually shorter than the leaves, loose, the limb subacute. Capitulum subglobose, 12–19 mm wide, white speckled with darker bracts; involucral bracts large and scarious, as wide as the capitulum, pallid, flushed grey or blackish, ovate to ovate-oblong, 4.5–6 mm long, acute, finally reflexed; floral bracts narrowly oblong, flushed grey, concave, densely white-villous above the middle, long-acuminate; receptacle thinly hairy. Flowers trimerous, 3.5–4.5 mm long. Male flowers: sepals grey, unequally joined below or up to the middle, the two lateral lightly keeled, the third concave, white-pilose across the tips; petals ligular, exserted from calyx, glandular and white-villous, dimorphic, the longest 2.1–3.5 mm long; anthers black. Female flowers: sepals oblong, grey, deeply concave, thickened and spongy along the midline and sometimes expanded into a narrow wing, villous along the lower margins inside, white-pilose at the tip; petals and ovary sessile; petals oblong-spathulate, glandular and white-villous upwards, the mid petal longer; seeds ellipsoid to broadly ellipsoid, 0.8–1 mm long, mid- to reddish-brown.

Malawi. N: Chitipa Dist., N end of Nyika Plateau, W foot of Nganda, 2350 m, 28.vii.1972, *Brummitt, Munthali & Synge* WC54 (K, LISC, SRGH).

Also in Kenya and Tanzania. Swampy ground and wet streamsides in montane grassland; c.2300–2500 m.

Conservation notes: Although widely distributed in Africa, apparently restricted to the Nyika Plateau within the Flora area, where it is Near Threatened.

E. mesanthemoides appears to be almost restricted to the Uluguru Mts in Tanzania and the Aberdares in Kenya, with an outlying population on the Nyika Plateau. However, it is so similar in appearance and habitat requirements to the better known *E. schimperi* that it may have been overlooked. The pointed involucral bracts are larger and thinner than in *E. schimperi*, radiating across the full width of the capitulum base, and the leaves are more tapering, especially in larger specimens. The female sepals of *E. schimperi* are never villous inside.

18. **Eriocaulon elegantulum** Engl., Pflanzenw. Ost.-Afr. **C**: 133 (1895). —N.E. Brown in F.T.A. **8**: 254 (1901). —Phillips in F.T.E.A., Eriocaulaceae: 17 (1997); in Kirkia **17**: 33 (1998). Type: Tanzania, Tanga Dist., Duga, vii.1893, *Holst* 3181 (K, HBG).

Slender tufted annual. Leaves linear, 2–7.5 cm long, 1.5–6 mm wide, thin, acute. Scapes up to about 50, often slightly flexuous, 7–25 cm high, 4-ribbed, the ribs forming narrow wings; sheaths about as long as the leaves, the limb with a delicate translucent acute tip, soon splitting. Capitulum globose, 3–5 mm wide, greyish-white, the pale female petals showing among the white-hairy floral bracts; involucral bracts shorter than the capitulum width, 4–5 spaced in one circlet (best seen in very young, still hemispherical capitula), ovate-oblong, obtuse, translucent, crumpled and obscured by the expanding flowers and apparently absent at maturity; floral bracts narrowly oblong-cuneate, black, scarious, concave, densely white-pilose upwards, acute; receptacle pilose, often thinly. Flowers trimerous, 1–1.2 mm long. Male flowers: calyx spathate, black, free sepal tips obtuse, white-pilose; petals small and included within calyx, white-pilose, glands very small; anthers black. Female flowers: sepals black, narrowly oblong, one slightly to definitely smaller, concave, lightly keeled and narrowly winged above the middle, white-hairy towards the acute tip; petals and ovary subsessile; petals narrowly oblanceolate-oblong, subequal, pallid, white-hairy around the tips, glands small, often absent on the median petal. Seeds broadly ellipsoid, 0.3 mm long, yellowish-brown, translucent, white-reticulate.

Zimbabwe. N: Gokwe Dist., Sengwa Research Station, 16.iv.1969, *Jacobsen* 622 (K, SRGH). S: Mwenezi Dist., Malangwe R., SW Mateke hills, 6.v.1958, *Drummond* 5656 (K, PRE, SRGH). **Malawi**. S: Kasupe Dist., Liwonde Nat. Park, near Mvuu camp, 14.iv.1985, *Dudley* 1526 (K). **Mozambique**. N: Meconta Dist., 51 km E of Camuana, 21.v.1961, *Leach & Rutherford-Smith* 10953 (LISC, SRGH). Z: Mocuba Dist., Mocuba, 20.iv.1948, *Faulkner* 256 (K).

Also S Sudan, Kenya, Tanzania (incl. Zanzibar & Pemba), Cameroon, Ghana and S Nigeria. Seasonal ponds, ditches, drainage lines and swampy grass areas, usually on sand, occasionally in rice fields; sea level to 600 m.

Conservation notes: Widely distributed; not threatened.

E. elegantulum can be difficult to distinguish from *E. transvaalicum* subsp. *hanningtonii*, another annual of similar appearance with thin linear leaves and greyish capitula on slender scapes. *E. transvaalicum* differs in its 5-ribbed unwinged scapes, involucral bracts in more than one series which remain visible in the mature capitulum, and larger seeds (0.5 mm).

19. **Eriocaulon bongense** Ruhland in Bot. Jahrb. Syst. **27**: 75 (1899). —N.E. Brown in F.T.A. **8**: 246 (1901). —Ruhland in Engler, Pflanzenr. **13**: 100 (1903). — Phillips in F.T.E.A., Eriocaulaceae: 18 (1997); in Kirkia **17**: 33 (1998). Type: Sudan, Bongoland, Bulu stream near Sabbi, 3.xii.1869, *Schweinfurth* 2722 (K, P). FIGURE 13.4.**3**.

Rosulate herb, fairly robust, probably annual. Leaves ensiform, 3–6 cm long, 4–7 mm wide, spongy, subacute. Scapes 1–16, flexuous, 10–35 cm high, 5–7-ribbed; sheaths exceeding the leaves, slightly inflated, the mouth deeply obliquely slit with an acute limb. Capitulum globose to conical, 5–8 mm wide, shaggy, shining pale silvery-grey with yellowish involucral bracts; involucral bracts forming a crateriform base to the capitulum, scarious with a leathery lower part, obovate, broadly rounded with small central cusp, reflexed at maturity; floral bracts oblanceolate to cuneate-oblong, scarious, glabrous, pallid flushed grey, incurving, the tip extended into a long slender cusp; receptacle pilose. Flowers trimerous, 2–2.5 mm long, pedicellate. Male flowers: sepals free, linear, subequal or one slightly smaller, pallid to grey, glabrous; petals tiny, glabrous or white-pilose, eglandular; anthers black. Female flowers: sepals narrowly linear, slightly geniculate, subequal or one noticeably smaller, pallid or grey, glabrous, acuminate; petals and ovary raised on a stipe; petals linear, white, subequal or the odd petal up to a third larger, eglandular, white-pilose at the tips. Seeds ellipsoid, 0.35 mm long, brown, white papillose-reticulate.

Zambia. W: Kasempa Dist., Mufumbwe (Chizera), 100 km WNW of Kasempa, 7.vii.1963, *Robinson* 5552 (K, SRGH). **Mozambique**. N: Mecula Dist., Niassa Game Reserve, Mbatamila staff camp, 12°11.01'S 37°33.01'E, 470 m, 6.vi.2003, *Timberlake et al.* 4892 (K, LMU).

West Africa eastwards to Sudan; also in S Tanzania. Marshy grassland, wet flushes and in rice fields; 450–1200 m.

Conservation notes: Although widely distributed in Africa, apparently localised within the Flora area; Near Threatened.

A distinctive species on account of the rather shaggy-looking pale capitula and the very narrow sepals and petals intermingling with the slender cusps of the floral bracts. The main centre of distribution is in West Africa, where the capitula are usually whitish-yellow or light silvery-grey. The few collections from Tanzania and Zambia have darker grey capitula.

20. **Eriocaulon abyssinicum** Hochst. in Flora **28**: 341 (1845). —N.E. Brown in Fl. Cap. **7**: 53 (1897); in F.T.A. **8**: 257 (1901). —Ruhland in Engler, Pflanzenr. **13**: 282 (1903). —Hess in Ber. Schweiz. Bot. Ges. **65**: 165 (1955). —Obermeyer in F.S.A. **4**(2): 11 (1985). —Phillips in F.T.E.A., Eriocaulaceae: 19 (1997); in Kirkia **17**: 34 (1998). Type: Ethiopia, Shire, 10.x.1840, *Schimper* 1944 (TUB holotype, K, G). FIGURE 13.4.**4**.

Eriocaulon gilgianum Ruhland in Bot. Jahrb. Syst. **27**: 84 (Apr. 1899). —Obermeyer in F.S.A. **4**(2): 14 (1985). Type: Angola, Huíla, *Antunes* 168a (B holotype).

Eriocaulon ciliisepalum Rendle in Hiern, Cat. Afr. Pl. Welw. **2**: 98 (May 1899). Type: Angola, Huíla, near Lopollo, *Welwitsch* 2445; Morro de Lopollo, *Welwitsch* 2445b (both BM syntypes, K).

Eriocaulon subulatum N.E. Br. in F.T.A. **8**: 255 (1901). Type: Zimbabwe, Zambezi R., island at Victoria Falls, 1860, *Kirk* s.n. (K holotype).

Small rosulate annual. Leaves few, light green, subulate, relatively long, exceeding the sheaths and often intermingling with the lower capitula, up to 5 cm long, 0.8–1.7 mm wide, septate, tapering to a fine acuminate tip. Scapes very slender, several to numerous and radiating to form a dome-shaped mound, 2–12 cm high, 3–4-ribbed; sheaths shorter than the leaves, mouth oblique with entire subacute limb. Capitulum globose to ovoid, 2–3 mm wide, grey to blackish with paler involucral bracts, loose and untidy with prominent stigmas; involucral bracts as wide as the capitulum, pallid, scarious, elliptic to narrowly ovate, subacute, not reflexing;

floral bracts oblanceolate-oblong, concave, subacute to shortly acuminate, scarious, flushed grey with the upper usually darker; receptacle glabrous to villous. Flowers trimerous, c.1 mm long. Male flowers: calyx spathate, grey, sepals joined below middle, the free tips triangular, glabrous, acute; petals rudimentary, glabrous; anthers black. Female flowers: sepals variable, subequal, usually flushed grey, narrowly lanceolate and slightly concave, varying to broader and deeply concave but not keeled, acute or acuminate, margins glabrous with a few short hairs, or occasionally conspicuously ciliate; petals subequal, linear, glabrous, vestigial glands usually present at least on the two smaller, sometimes emarginate, especially the larger. Seeds ellipsoid to broadly ellipsoid, 0.35 mm long, brown, glossy, ± smooth with a faint reticulation.

Botswana. SE: Gaborone, N of Kgale Siding, 8.iv.1978, *Hansen* 3400 (K, SRGH). **Zambia.** N: Mbala Dist., Kalambo R., Sansia Falls, 8.v.1961, *Richards* 15133 (SRGH). C: Lusaka Dist., 14 km WNW of Lusaka, Yieldingtree Farm, 9.v.1993, *Bingham* 9338 (K). E: Petauke Dist., 32 km E of Kachalolo, 24.iii.55, *Exell, Mendonça & Wild* 1169 (BM, LISC, SRGH). S: Livingstone Dist., Victoria Falls, 26.v.1960, *Fanshawe* 5695 (K, SRGH). **Zimbabwe.** N: Guruve Dist., Nyarasuswe, 14.v.1962, *Wild* 5740 (K, SRGH). W: Matobo Dist., Besna Kobila Farm, iv.1959, *Miller* 5897 (K, SRGH). C: Harare Dist., Mukuvisi (Makabusi) R., 24.iv.1948, *Wild* 2510 (K, SRGH). **Malawi.** C: Dedza Dist., Mua–Livulezi Forest Reserve, 19.iii.1955, *Exell, Mendonça & Wild* 1069 (LISC, SRGH).

Tropical Africa from Ethiopia southwards through East Africa to N Namibia and South Africa, also in N Nigeria and Cameroon. Pioneer, on shallow muddy soil overlying rock outcrops, usually with flowing water, in the mist zone of waterfalls, and in seasonal wet flushes and ponds in grassland. Locally abundant in association with other small ephemerals including *E. cinereum* and *E. mutatum* and *Xyris* spp.; 650–1400 m.

Conservation notes: Widely distributed; not threatened.

E. abyssinicum is a widely distributed and variable small pioneer species, recognised by its numerous slender scapes topped by small dark capitula with paler involucral bracts, and surrounded by a rosette of rather long, narrow, finely tapering leaves. The main source of variation lies in the degree of hairiness of the receptacle and female sepal margins. It has traditionally been separated from other species listed in synonymy by the possession of a glabrous receptacle and sepals. In fact, a few hairs are usually present in one or both locations and are often more numerous and obvious. Whilst plants from the southern part of the range tend to be hairier, glabrous plants occur as far south as KwaZulu-Natal and hairy ones north to Uganda. A division into species on this character is not tenable. *E. abyssinicum* intergrades with *E. welwitschii* (see note under *E. welwitschii*).

E. abyssinicum is frequently confused with *E. mutatum* and *E. cinereum*, all three being very similar small dark-headed annuals which sometimes grow together in open situations with other small pioneer species. *E. cinereum* can be distinguished by its white anthers and oblong obtuse floral bracts, and *E. mutatum* has shorter, broader leaves, less obvious involucral bracts and broader, less scarious floral bracts. On dissection, it can be immediately distinguished from *E. abyssinicum* by its dimerous flowers.

21. **Eriocaulon welwitschii** Rendle in Hiern, Cat. Afr. Pl. Welw. **2**: 97 (1899). —N.E. Brown in F.T.A. **8**: 249 (1901). —Ruhland in Engler, Pflanzenr. **13**: 102 (1903). —Hess in Ber. Schweiz. Bot. Ges. **65**: 270 (1955). —Obermeyer in F.S.A. 4(2): 13 (1985). —Phillips in F.T.E.A., Eriocaulaceae: 21 (1997); in Kirkia **17**: 35 (1998). Type: Angola, Pungo Andongo, between Lombe and Candumba, iii.1857, *Welwitsch* 2441 (BM holotype, B, K, M).

Eriocaulon welwitschii var. *pygmaeum* Rendle in Hiern, Cat. Afr. Pl. Welw. **2**: 98 (1899). Type: Angola, Huíla, near Lopollo, v.1860, *Welwitsch* 2444 (BM holotype, B, K, M).

Eriocaulon aristatum H. Hess in Ber. Schweiz. Bot. Ges. **65**: 163 (1955). Type: Angola, Humpata Dist., Serra da Chela, 1850 m, 15.v.1952, *Hess* 52/1755a (Z holotype, BR).

Small rosulate annual. Leaves pale green, needle-like, 1–2(5) cm long, 0.2–1 mm wide, septate, soon dying back. Scapes very slender, numerous, radiating to form a dome-shaped mound, 1–12 cm high, 3–4-ribbed; sheaths shorter than the leaves, scarious, obliquely slit. Capitulum hemispherical to dome-shaped, spiny, 2.5–4 mm wide, scarious, whitish to grey, the upper bracts often darker grey; involucral bracts as wide as the capitulum and radiating across its base, lanceolate with sharply acuminate tips, pale; floral bracts narrowly lanceolate, acuminate-aristate, flushed pale grey; receptacle columnar, glabrous to villous. Flowers trimerous, pallid or flushed pale grey, 0.8–1 mm long. Male flowers: calyx spathate, sepals free or ± united, the tips acuminate; petals tiny, glandular; anthers black. Female flowers: sepals equal, linear-lanceolate, concave, acuminate, margins glabrous or with a few short hairs; petals linear, tipped with a rudimentary gland. Seeds elliptic, 0.3 mm long, brown, glossy, faintly reticulate.

Botswana. N: Okavango Delta, bay on E side of Xere Is., 19°06'S 23°06.5'E, *P.A. Smith* 634 (LISC, SRGH). **Zambia**. W: Mufulira, 15.v.1934, *QVM* 8120 (SRGH). S: Choma Dist., Muckle Neuk, 19 km N of Choma, 27.ii.1954, *Robinson* 577 (K, SRGH). **Zimbabwe**. W: Hwange Dist., 80 km W of Main Camp, Shumba Pans, Hwange Nat. Park, 18.iv.1972, *Gibbs Russell* 1668 in part (mixed with *E. cinereum*) (K, SRGH). C: Harare (Salisbury), 25.iv.1931, *Brain* 3629 (SRGH). S: Mwenezi Dist., Malangwe R., SW Mateke Hills, 6.v.1958, *Drummond* 5597 (K, LISC, SRGH).

Also in S Tanzania, Angola and Namibia. Wet sandy ground in open places; 600–1300 m.

Conservation notes: Widely distributed; not threatened.

E. welwitschii and *E. abyssinicum* represent different facets of a single intergrading complex, and consequently intermediates occur between them (e.g. *Robinson* 3694 from Zambia). The receptacle of both taxa varies from glabrous to villous and, like *E. abyssinicum*, the female sepals of *E. welwitschii* may be either glabrous or ciliate. *E. welwitschii* is distinguished by the combination of generally paler and narrower bracts and perianth parts, the tips of which are very narrowly pointed, imparting a spiny appearance to the capitulum. The seed also tends to be a little smaller and more narrowly ellipsoid than in *E. abyssinicum*.

22. **Eriocaulon wildii** S.M. Phillips in Kew Bull. **52**: 64 (1997); in Kirkia **17**: 36 (1998). Type: Zimbabwe, Nyanga Dist., van Niekerk Ruins, Ngarawe R., 2.viii.1950, *Wild* 3512 (K holotype, LISC, SRGH).

Small rosulate annual. Leaves subulate, 1–2 cm long, c.1 mm wide, spongy, subacute. Scapes up to c.10, stout, stiff, 4-ribbed; sheaths shorter than the leaves, the mouth inflated, papery, soon splitting. Capitulum hemispherical, 5 mm wide; involucral bracts narrowly lanceolate-oblong, much paler than the the floral bracts, as wide as the capitulum and spreading in 2–3 series, scarious, subacute; floral bracts narrowly lanceolate, dark grey, glabrous, acute; receptacle glabrous; flowers trimerous, 1.2–1.5 mm long, glabrous. Male flowers: sepals united into an open spathe with free, acute tips, blackish; petals seated on a linear stipe, small, glabrous, tipped by a black gland; anthers 6, black. Female flowers: sepals subequal, ovate, concave, a spongy, swollen wing on the centre back, wing about as wide as the sepal body, its margin sometimes toothed, sepal translucent below, flushed grey towards the narrowly acute tip; petals subequal, linear-spathulate, tips entire with a black gland. Seeds ellipsoid, 0.4 mm long, light brown, glossy, almost smooth.

Zimbabwe. E: Nyanga Dist., van Niekerk Ruins, Ngarawe R., in wet sand over granite, 1200 m, 2.viii.1950, *Wild* 3512 (K, LISC, SRGH); Mutare Dist., 24 km from Mutare (Umtali) on Chimanimani (Melsetter) road, 4.vii.1948, *Chase* 886 (K, SRGH).

Known only from the Eastern Highlands of Zimbabwe, on wet sand by rivers; c.1200 m.

Conservation notes: Apparently endemic to E Zimbabwe. Restricted distribution, although partly in protected areas; probably Near Threatened.

E. wildii is most closely related to *E. aethiopicum* S.M. Phillips from S Ethiopia, which has shorter ovate involucral bracts, a pilose receptacle, and a thin scattering of short white hairs on the floral bracts and flowers. Within the Flora area it is most likely to be confused with *E. welwitschii*, which has similar narrow pointed bracts, but never has winged sepals.

23. **Eriocaulon infaustum** N.E. Br. in F.T.A. **8**: 253 (1901). —Phillips in Kirkia **17**: 36 (1998). Type: Mozambique, Quelimane, viii.1887, *Scott* s.n. (K holotype).

Tufted annual. Leaves linear, 10–13 cm long, 3.5–4 mm wide, exceeding sheaths and equalling the shortest scapes, thin, septate, tapering to an acute tip. Scapes up to 20, fairly stout, 15–22 cm high, sharply 5-ribbed; sheaths obliquely slit with acute limb. Capitulum subglobose, 5.5–6 mm wide, greyish-white, female petals visible among the bracts; outermost involucral bracts radiating beyond the periphery of the young capitulum, linear-oblong, obtuse, 3.3 mm long, scarious, yellowish, crumpling and reflexing at maturity, the inner similar but shorter and broader; floral bracts angulate-oblanceolate, blackish with paler margins, densely white-pilose towards the sharply acute tip; receptacle pilose. Flowers 1.5–1.6 mm long, pedicellate. Male flowers: sepals united into a blackish flattened spathe with a dentate, slightly 3-lobed upper margin, glabrous; petals tiny, included within calyx, glandular and densely white-pilose; anthers black. Female flowers: sepals subequal, oblong to boat-shaped, weakly keeled, glabrous or a few white papillae near the tips, laterals falcate, the median straight, unwinged but spongy at the centre keel, acute; petals raised on a short stipe, subequal, narrowly oblong-spathulate, glandular and white-pilose at the tips, a few scattered white hairs on the inner blade. Seeds ellipsoid, 0.5 mm long, reddish brown, glistening, closely transversely white papillose-striate.

Mozambique. Z: Quelimane (Quilimane), viii.1887, *Scott* s.n. (K).
Known only from the type specimen, collected in rice fields; c.10 m.
Conservation notes: Known only from the type, collected over 100 years ago; Data Deficient, possibly Extinct.

24. **Eriocaulon matopense** Rendle in J. Linn. Soc., Bot. **27**: 475 (1906). —Phillips in Kirkia **17**: 37 (1998). Type: Zimbabwe, Matobo Hills, bog near "View", ix.1905, *Gibbs* 201 (BM holotype, K).

Eriocaulon dehniae H.E. Hess in Ber. Schweiz. Bot. Ges. **67**: 84 (1957). Type: Zimbabwe, Rusape, 22.ii.1954, *Dehn* 1071 (Z holotype, K).

Perennial from a short rootstock. Leaves many in a basal tuft, erect, linear, 2–6 cm long, 1–2 mm wide, tapering to a narrow tip with a circular pore on the upper side. Scapes 2–20, up to 30 cm high, 5–7-ribbed; sheaths equalling or slightly longer than the leaves, the mouth obliquely slit with obtuse limb, lacerate. Capitulum hemispherical becoming dome-shaped, 4.5–6.5 mm wide, pale creamy-grey, bracts imbricate with white-hairy tips of the flowers visible among them; involucral bracts in several series, equalling the capitulum and forming a flat or slightly intruded base, straw-coloured to golden, obovate to obovate-oblong, coriaceous, glabrous; floral bracts obovate to angulate, firmly membranous, pale or flushed grey, white-hairy on the central back, often thinly, tip broadly triangular to cuspidate; receptacle pilose. Flowers trimerous, 1.5–2 mm long. Male flowers: sepals free, oblong, white with grey tips, two lateral boat-shaped with a concave keel, the third flat, all thinly white-pilose near the obliquely truncate tip; petals raised on a columnar stipe as long as the petals, glandular, unequal, the longest c.0.6 mm, white-hairy; anthers black. Female flowers: sepals resembling the male, subequal, oblanceolate, two lateral with a thickened or narrowly winged keel; petals subequal, narrowly oblanceolate, white-hairy on the inner blade upwards, often with a few hairs also on the outer face, glands variable, sometimes vestigial or absent on one or more petals. Seeds plumply ellipsoid, 0.4 mm long, brown with white papillae, mostly in transverse rows but sometimes reticulate.

Zimbabwe. W: Matobo Dist., Besna Kobila Farm, x.1956, *Miller* 3675 (K, SRGH). C: Harare Dist., Shamva road, Mansala R., 24.x.1960, *Rutherford-Smith* 341 (K, SRGH); Makoni Dist., Rusape, 22.ii.1954, *Dehn* 1071 (K, LISC, Z); Goromonzi Dist., Mermaid's Pool, 5.vii.1934, *Gilliland* 620 (BM). E: Nyanga (Inyanga), 25.xi.1930, *Fries, Norlindh & Weimarck* 3225 (SRGH).

Endemic to the central watershed of Zimbabwe; unknown elsewhere. Marshy ground and shallow water of streams; 1300–2000 m.

Conservation notes: Distributed across a fairly broad but agricultural area; probably not threatened.

25. **Eriocaulon zambesiense** Ruhland in Bot. Jahrb. Syst. **27**: 75 (1899). —Phillips in Kirkia **17**: 38 (1998). Types: Malawi, Shire Highlands, vii.1885, *Buchanan* s.n. & Zomba Mt, xii.1896, *Whyte* s.n. (both B syntypes, K).

Rosulate perennial from a short rhizome. Leaves linear, 5–12 cm long, 3–4 mm wide, spongy, subacute. Scapes 5–15, 10–35 cm high, 5–6-ribbed; sheaths equalling the leaves, obliquely slit, the limb acute. Capitulum subglobose, slightly wider than long, 5.5–7 mm wide, black and white, the white-hairy petal tips intermingling with the darker bracts, sometimes viviparous; involucral bracts small, straw-coloured or flushed grey, cartilaginous, oblong, rounded, glabrous, reflexed at maturity; floral bracts narrowly cuneate or oblanceolate, grey, white-pilose towards the acute tip; receptacle villous. Flowers trimerous, 2.1–2.5 mm long. Male flowers: sepals blackish, oblong, keeled, unequally joined below into a spathe, the tips broadly rounded, white-pilose; petals shortly exserted from calyx, slightly unequal, white-pilose at tips, the longest 0.7–1 mm long; anthers black. Female flowers wedge-shaped, strongly trigonous; sepals boat-shaped, blackish, the median narrower, all strongly keeled, deeply concave and gibbous, keel with a spongy wing, white-pilose around the tip, margins villous along the inside with tubercle-based translucent hairs; petals and ovary sessile; petals narrowly oblanceolate, slightly unequal, white-pilose at tips, hairy on the inner blade. Seeds ellipsoid, 0.7 mm long, yellow or brown, transversely white papillose-striate.

Malawi. S: Zomba Dist., Zomba Plateau between Chiradzula and Malumbe, 10.x.1971, *Jordan* 5004 (K). **Mozambique**. Z: Gurué Dist., Mt Namuli, Muretha plateau, 19.xi.2007, *Mphamba* 25 (K, LMA).

Also in Burundi. Marshy grassland and wet mud of streamsides; c.1800 m.

Conservation note: Within the Flora area only known from mountains in S Malawi (Zomba Mt and nearby) and N Mozambique (Mt Namuli); probably Vulnerable.

The female flowers, with their hairy margined sepals with a spongy winged keel, are very reminiscent of those of *E. sonderianum* from Mt Mulanje which, however, has much thinner, tapering, finely pointed leaves and usually only a single scape per rosette.

The populations in S Malawi and Burundi differ slightly and a separation at subspecific level may be justified when the species is better known. The description above refers to the typical form in Malawi, which is remarkable for its strongly trigonous female flowers with the median sepal similarly keeled and winged to the laterals. In the few specimens from Burundi seen by the author the flowers are slightly smaller (1.7 mm) and bilaterally compressed rather than trigonous. This difference in shape arises from the much narrower, oblong, unwinged median sepal in the Burundi plants.

26. **Eriocaulon inyangense** Arw. in Bot. Not. **1934**: 83 (1934). —Phillips in F.T.E.A., Eriocaulaceae: 22 (1997); in Kirkia **17**: 38 (1998). Type: Zimbabwe, Nyanga, Niarawe stream, 1650 m, fl.& fr. 31.x.1930, *Fries, Norlindh & Wiemarck* 2478 (LD holotype, BM, SRGH).

Eriocaulum katangaense Kimpouni in Fragm. Fl. Geobot. **39**: 337 (1995). Type: Congo, Katanga, 6 km E of Kasumbalesa, Lumbumbashi–Sakania road, 8.viii.1958, *Schmitz* 6131 (BR holotype).

Rosulate herb, probably perennial. Leaves in a dense tuft, linear-subulate to lanceolate, 2–5.5 cm long, 2–3(5) mm wide, outwardly curving, spongy, acute. Scapes up to 25, slender, 7–35 cm high, 5–6(7)-ribbed; sheaths equalling the leaves, obliquely slit, the scarious limb splitting into 2–3 lobes. Capitulum dirty white, 4–5 mm wide, initially globose with a flat base, becoming slightly elongated and dome-shaped with intruded base at maturity; involucral bracts shorter than the capitulum width, obovate-oblong, rounded, light brown, coriaceous at base becoming scarious and greyish upwards; floral bracts broadly spathulate, cuspidate, incurving, scarious, pale grey-brown, thinly white-pilose on the widest part; receptacle columnar, villous. Flowers 1.3–1.8 mm long, trimerous, pedicellate. Male flowers: sepals partially united into a spathe, laterals concave, the median flat, grey, tips obtuse and white-pilose; petals subequal, glandular and white-pilose, the largest 0.5–0.7 mm long, shortly exserted from calyx; anthers black. Female flowers: sepals subequal, grey, deeply concave, lightly keeled, narrowly winged or at least spongily thickened on centre keel, white-pilose with short stout hairs on back and margins towards the tip, margins usually pilose inside with longer translucent hairs, tip subacute; petals subequal, linear-oblanceolate, glandular, white-pilose at tips, pilose on the inner blade with longer translucent hairs. Seeds elliptic, 0.5 mm long, brown, transversely white papillose-striate.

Zambia. N: Mbala Dist., Mbala (Abercorn), marsh, left side of pans, 11.ii.1957, *Richards* 8158 (K). C: Serenje, 2.xi.1972, *Fanshawe* 11655 (K). **Zimbabwe**. E: Chimanimani Dist., Martin Forest Reserve, 16.xi.1967, *Mavi* 664 (K, LISC, SRGH).

Also in Kenya, Tanzania and Congo (Katanga). Marshy grassland and streamsides; 1400–1800 m.

Conservation notes: Widely distributed; not threatened.

E. inyangense has a typical appearance of small, greyish, slightly elongate capitula on slender scapes rising above a basal tuft of often outwardly curving leaves. It is closely related to *E. zambesiense*, which has a similar but rather more robust perennial habit. However, it can be distinguished from *E. inyangense* by its slightly larger, more obviously white-hairy capitulum which is never longer than wide, and especially by the much broader spongy wing to the female sepals. *E. inyangense* has most frequently been confused with the tall slender form of *E. transvaalicum* subsp. *hanningtonii*, but this has flatter, thinner leaves up to 5 mm wide and seldom occurs above 1500 m.

The type specimen has rather stouter scapes than usual with 7 ribs, but otherwise matches the rest of the material well. The number of ribs in the scape of *Eriocaulon* species often varies within narrow limits and is not in itself a character of specific significance.

E. katangaense is probably this species, but the type is very immature.

27. **Eriocaulon transvaalicum** N.E. Br. in Fl. Cap. **7**: 54 (1897). —Ruhland in Engler, Pflanzenr. **13**: 81 (1903). —Hess in Ber. Schweiz. Bot. Ges. **65**: 155 (1955). —Obermeyer in F.S.A. **4**(2): 17 (1985). —Phillips in F.T.E.A., Eriocaulaceae: 23 (1997); in Kirkia **17**: 39 (1998). Type: South Africa, former Transvaal, Bosveld, near Boekenhoutskloof, 1875, *Rehmann* 4787 (K holotype, BM, Z).

Eriocaulum schweickerdtii Moldenke in Phytologia **3**: 416 (1951). Type: Zimbabwe, Mutare Dist., Nyamshatu R., vii.1948, *Fisher & Schweickerdt* 234 (SRGH holotype, PRE).

Rosulate annual or short-lived perennial, sometimes forming colonies. Leaves broadly linear to sword-shaped, 2–10 cm long, 2–6.5 mm wide, flat, thin, septate, subacute. Scapes up to 30, usually stiff, 8–35 cm high, 5-ribbed; sheaths a little shorter than the leaves, obliquely slit, the limb obtuse or splitting into 3–4 dry brownish teeth. Capitulum globose, shiny black, greyish-white or infrequently brownish-grey, (3)5–9 mm wide, the dark female petal-tips intermingling with the bracts; involucral bracts as wide as or shorter than the capitulum width, scarious, usually paler than rest of the capitulum, oblong to obovate, obtuse, glabrous, reflexed at maturity; floral bracts spathulate to angular-oblanceolate, scarious, dark grey, acute, acuminate or cuspidate, varying from ± glabrous to pilose with short spaced white hairs on the upper back; receptacle villous. Flowers trimerous, 1.5–2 mm long. Male flowers: sepals black, joined into a spathe deeply split on one side, lobes free at the tips, glabrous or thinly white-hairy above, the upper margin often denticulate; petals reduced, included within calyx, glands small or absent; anthers black. Female flowers: sepals subequal, dark grey, variable, boat-shaped, narrowly lanceolate-falcate to ovate, winged on the centre keel or at least slightly thickened there, narrowed to the base, acute to sharply acuminate, glabrous or thinly white-pilose upwards, the third narrower and often not keeled; petals and ovary subsessile to stipitate; petals unequal, linear with dark tips, glabrous or white-pilose at tip, infrequently some longer translucent hairs also on the upper blade, glands vestigial or absent. Seeds ellipsoid, 0.4–0.6 mm long, yellowish-brown or infrequently reddish-brown, transversely white papillose-striate.

Northwards through East Africa to Ethiopia, and in Angola, Namibia and South Africa (Limpopo, KwaZulu-Natal and Free State). In mud of shallow water of streams, ditches and wet flushes in grassland, the leaves often submerged.

A widespread and variable annual of wet muddy places, recognised by its broad, thin leaves and many black heads on stiff, relatively stout scapes. The female sepals vary considerably in width and shape, depending on how broad the wing is at the centre of the keel (often merely an indistinct thickening), and the degree to which the tips are acuminately extended. The female petals are frequently glabrous and eglandular, or may bear just a few short white hairs near the tip. There are also microscopic differences to the seed surface patterning. These differences are discussed by Phillips (Kew Bull. **51**: 625–647, 1996), who subdivides the species on the basis of floral and seed characters into 4 subspecies, keyed out below. However, not all specimens will fit neatly into one or other of the subspecies, especially those from Zimbabwe where distributions overlap.

The type of *E. schweickerdtii* is wrongly cited in the protologue as being from Natal, but is in fact from Zimbabwe. It lacks seed and is not easily assigned to subspecies.

1. Capitulum black, ± glabrous · 2
– Capitulum pilose with stout thick white hairs · 3
2. Sepals of female flowers ovate, deeply concave, margins smooth, tip acute; seed surface ruminate · subsp. *transvaalicum*
– Sepals of female flowers narrowly falcate, margins denticulate above, tip cuspidate; seed surface tuberculate · · · · · · · · · · · · · · · · · · subsp. *dembianense*
3. Scapes stout, stiff · subsp. *tofieldifolium*
– Scapes slender, often flexuous · subsp. *hanningtonii*

Subsp. **transvaalicum**

Leaves 4–5 cm long, 4–7 mm wide; scapes up to 16 cm high, c.1 mm thick; capitulum shiny black, floral bracts acute, a few white hairs on the upper back; sepals of female flowers ovate, deeply concave, margins smooth, tip acute; seed surface ruminate or ruminate-tuberculate.

Botswana. SE: Lobatse, c.1 km S of Boschwelathon Kloof, 20.viii.1955, *McConnell* in GH 68899 (SRGH).

Also in N South Africa; 1000–1400 m.

Conservation notes: Restricted distribution in the Flora area, possibly Vulnerable.

This subspecies has the usual facies of shiny black capitula on stout, stiff scapes associated with the species, but has broader sepals and a slightly different seed-coat patterning from populations occurring further north.

Subsp. **dembianense** (Chiov.) S.M. Phillips in Kew Bull. **52**: 57 (1997). —Phillips in F.T.E.A., Eriocaulaceae: 23 (1997); in Kirkia **17**: 41 (1998). Types: Ethiopia, Dembia, near Mt Inceduba, Gondar, 26.viii.1909, *Chiovenda* 1651 (FT syntype, K) and Scinta valley near Asoso, *Chiovenda* 1912 & 2573 (both FT syntypes).

> *Eriocaulon dembianense* Chiov. in Ann. Bot. (Roma) **9**: 148 (1911).

Leaves sword-shaped, 2–10 cm long, 2–5 mm wide; scapes stiff, stout, 8–15(20) cm high, 0.6–0.8 mm thick, capitulum shiny black or infrequently brownish-grey, glabrous; floral bracts acuminate or cuspidate, usually ± glabrous, occasionally a few short white hairs on the back; sepals of female flowers narrowly falcate, the upper margins denticulate, narrowly winged, tip cuspidate; seed surface tuberculate.

Zambia. N: Mbala Dist., Chisungu stream, Chilongowelo, 24.ii.1952, *Richards* 1545 (K). W: Kitwe, 22.iv.1966, *Mutimushi* 1375 (K, SRGH).

The commonest form in N and E tropical Africa, extending into N Zambia. Also in Ethiopia, Uganda, Kenya, Tanzania and Congo. Dambos, seepages and ditches; 1200–1500 m.

Conservation notes: Widely distributed; not threatened.

Subsp. *dembianense* resembles subsp. *transvaalicum* in its black shiny capitula on stout, stiff scapes, but differs in details of sepal shape and microscopic seed surface patterning.

Subsp. **tofieldifolium** (Schinz) S.M. Phillips in Kew Bull. **52**: 58 (1997). —Phillips in F.T.E.A., Eriocaulaceae: 24 (1997); in Kirkia **17**: 41 (1998). Type: Namibia, Waterberg, *Dinter* 378 (Z holotype).

> *Eriocaulum tofieldifolium* Schinz in Bull. Herb. Boiss., sér.2, **1**: 779 (1901). —Hess in Ber. Schweiz. Bot. Ges. **65**: 266 (1955). —Friedrich-Holzhammer & Roessler in Merxmüller, Prodr. Fl. SW Afr., fam.159: 2 (1967).

Leaves sword-shaped, 4–7 cm long, 2.5–5 mm wide; scapes robust, stiffly upright, 10–20 cm high, c.1 mm thick; capitulum 6–9 mm wide, greyish-white; floral bracts white-pilose above the middle; sepals of female flowers variable, ovate to lanceolate, wing narrow to broad with a toothed margin.

Zimbabwe. C: Harare (Salisbury), 5.v.1931, *Brain* 4545 (K).

Also in N South Africa, N Namibia and Tanzania; c.1450 m.

Conservation note: Apparently very restricted within the Flora area, where it might be Vulnerable, but probably not threatened globally.

Robinson 5358 from Malawi (Mulanje Dist., 16 km NW of Likabula, 15.vi.1962 (SRGH)) is tentatively placed here, but has a somewhat reticulate seed coat pattern, perhaps through introgression from *E. selousii*.

Subsp. **hanningtonii** (N.E. Br.) S.M. Phillips in Kew Bull. **52**: 58 (1997). —Phillips in F.T.E.A., Eriocaulaceae: 24 (1997); in Kirkia **17**: 41 (1998). Type: Tanzania, Morogoro Dist., Kwa Chiropa, vi.1803, *Hannington* s.n. (K holotype).

> *Eriocaulum hanningtonii* N.E. Br. in F.T.A. **8**: 253 (1901). —Ruhland in Engler, Pflanzenr. **13**: 74 (1903).
> *Eriocaulum transvaalicum* N.E. Br. var. *hanningtonii* (N.E. Br.) Meikle in Kew Bull. **22**: 142 (1968); in F.W.T.A., ed.2 **3**: 63 (1968).

Leaves broadly linear, up to 10 cm long, (2)4–6.5 mm wide; scapes 9–35 cm high, 0.5 mm thick, often flexuous; capitulum 4–6.5 mm wide, greyish-white; floral bracts white-pilose on the upper back; sepals of female flowers usually narrowly oblong-falcate, very narrowly winged, glabrous or a few white hairs near the tips.

Zambia. C: Kabwe (Broken Hill), 17.viii.1964, *Mutimushi* 956 (K, SRGH). E: Lundazi Dist., near Kalindi, Lukusuzi Nat. Park, 4.vii.1971, *Sayer* 1267 (SRGH). **Zimbabwe**. C: Harare, Mukuvisi (Makabusi) R., 24.iv.1948, *Wild* 2511 (K, SRGH). **Malawi**. C: Kasungu Dist., Kasungu Game Res., 11.vii.1970, *Hall-Martin* 1744 (SRGH). **Mozambique**. Z: Mocuba Dist., Mocuba, vii.1948, *Faulkner* K22 (K, SRGH). MS: Cheringoma Plateau, 11.vii.1972, *Ward* 7806 (K).

Also in Ethiopia, Kenya and Tanzania. Marshy depressions and seepages over rocks; sea level–1400 m.

Conservation note: Widely distributed; not threatened.

Subsp. *hanningtonii* is distinguished by its slender, flexuous, often taller scapes and white-hairy capitula. It occurs mainly at low altitudes in eastern Africa.

28. **Eriocaulon selousii** S.M. Phillips in Kew Bull. **52**: 60 (1997). —Phillips in F.T.E.A., Eriocaulaceae: 25 (1997); in Kirkia **17**: 42 (1998). Type: Tanzania, Selous Game Reserve, c.8 km S of Ruaha camp, seepage in miombo valley, 275 m, 20.x.1975, *Vollesen* 2896 (K holotype, C).

Slender rosulate annual. Leaves broadly linear, 1.5–3.5 cm long, 1–3 mm wide, septate, subacute. Scapes up to 15, 6–17 cm high, much taller than the small basal rosette of leaves, 4–5-ribbed; sheaths equalling the leaves. Capitulum subglobose, 4–5 mm wide, brownish-grey to dark grey with a slight sheen, glabrous, bracts loose and untidy, the flowers visible between; involucral bracts as wide as the capitulum, yellowish-grey, scarious, narrowly obovate, obtuse and sometimes minutely denticulate, reflexed at maturity; floral bracts resembling the involucral but narrower, cuneate-oblanceolate, shallowly incurving, obtuse and minutely denticulate, the inner acute; receptacle villous. Flowers trimerous, 1.5–1.7 mm long, glabrous. Male flowers: sepals joined into a thinly scarious grey spathe with free tips, the spathe wide open, only the margins inturned, tips truncate-denticulate; petals very small, included within calyx, eglandular; anthers black. Female flowers: sepals oblong with truncate-denticulate tips, laterals lightly keeled, the median similar but slightly narrower; all petals longer than sepals, narrowly oblanceolate-oblong, eglandular, the median emarginate, laterals bidentate; ovary sessile. Seeds ellipsoid, 0.4 mm long, brown with white-papillose reticulations.

Malawi. S: Blantyre Dist., Limbe, Tung Station, 17.viii.1950, *Jackson* 108 (K); Chikwawa Dist., Chikwawa (Shibisa) to Tshinsinze (Tshinmuze), 600–1200 m, ix.1859, *Kirk* s.n. (K).

Also in Tanzania. Seepages in *Brachystegia* woodland; 600–1200 m.

Conservation notes: Restricted distribution within the Flora area; probably Near Threatened.

The type formed part of a mixed gathering with *E. bongense* Ruhl. *E. selousii* is a segregate from the variable *E. transvaalicum* complex, distinguished mainly by its reticulate seed. *Kirk* s.n. is intermediate, with a reticulate seed, but definite wings on the female sepals and some white hairs on the floral bracts.

29. **Eriocaulon afzelianum** Körn. in Linnaea **27**: 680 (1861). —N.E. Brown in F.T.A. **8**: 250 (1901). —Ruhland in Engler, Pflanzenr. **13**: 83 (1903). —Phillips in F.T.E.A., Eriocaulaceae: 26 (1997); in Kirkia **17**: 42 (1998). Type: Sierra Leone, *Afzelius* s.n. (B holotype, BM, S).

Eriocaulum kouroussense Lecomte in Bull. Soc. Bot. France **55**: 644 (1909). Types: Guinea, marsh of Kouroussa, xii.1900, *Pobéguin* 615 (P syntype, K) and 616 (P syntype).

Rosulate herb, probably annual. Leaves linear to linear-lanceolate, up to 7 cm long, 1.5–4 mm wide, thin, septate, acute. Scapes up to 25, very slender, 7–35 cm high, sharply 4–5-ribbed; sheaths ± equalling the leaves, deeply obliquely slit with long acute limb. Capitulum 4–6 mm wide, subglobose to slightly dome-shaped, greyish with mealy white pubescence or sometimes dark grey, the flowers hidden except for the largest female petal at maturity; involucral bracts yellowish, broadly oblong to lanceolate-oblong with rounded tips, firm, often thinly mealy-pubescent, slightly smaller than the capitulum width and appressed to it in several series; floral bracts neatly imbricate, narrowly cuneate-spathulate, scarious, uniformly grey or with pallid tips, incurved, white-pilose on the widest part, sharply acute; receptacle villous. Flowers trimerous, shortly pedicellate. Male flowers 1.6–2 mm long; sepals partially joined into a thin scarious spathe open on one side or sometimes almost free, oblong, the laterals concave, all white-pilose towards the obliquely truncate tips; petals included within calyx, median larger than the laterals, eglandular, white-pilose; anthers blackish. Female flowers 2–2.5 mm long: sepals subequal, grey, narrowly oblanceolate-oblong, obtuse, thinly white-pilose towards the tips, the lateral two falcate, very lightly keeled, the median almost flat; petals raised on a short stipe, unequal, spongy, eglandular, white-ciliate on the upper margins, white-hairy also on both sides of the blade especially on the larger median petal which is longer than the sepals, elliptic with narrow base, the laterals smaller, narrowly oblanceolate; ovary blackish. Seeds subglobose, c.5 mm long, blackish, reticulate (reticulations finally white).

Zambia. N: Mbala Dist., Kali Dombo, near Kawimbi Mission, 1550 m, 6.v.1952, *Richards* 1634 (K); Mporokoso Dist., Nsama-Mporokoso road, 16.iv.1957, *Richards* 9274 (K). W: Mwinilunga, Kalenda Plain, Matonchi, 16.iv.1960, *Robinson* 3619 (K). **Zimbabwe**. C: Chegutu (Hartley), Poole Farm, 3.v.1945, *Hornby* 2388 (SRGH). **Malawi**. N: Karonga Dist., near Kayelekera, 32 km WSW of Karonga, 4.vi.1989, *Brummitt* 18331 (K).

Also in W Africa from Senegal to Chad and Central African Republic and Tanzania. Swampy grassland and wet flushes on rocky hillsides; 900–1200 m.

Conservation notes: Widely distributed, not threatened.

E. afzelianum is a widespread annual of variable habit, but well-grown specimens are rather tall with flexuous scapes and thin cellular leaves often over 5 cm long. The southern population in the Flora area has darker grey capitula than West African specimens, but there are no other differences.

The capitula always have pale, appressed involucral bracts and regularly imbricated, acute floral bracts. The female flowers are characterised by narrow, subequal sepals and unequal, spongy, eglandular petals hairy on both sides. The ovary is a striking blackish colour, as are the reticulately patterned seeds.

30. **Eriocaulon stenophyllum** R.E. Fr., Wiss. Ergebn. Schwed. Rhod.-Kongo-Exped. **1**: 218, t.16/1–3 (1916). —Phillips in F.T.E.A., Eriocaulaceae: 25 (1997); in Kirkia **17**: 43 (1998). Type: Zambia, Kali, between Mansa and Lake Bangweulu, 17.ix.1911, *R.E. Fries* 634a (UPS lectotype, K).

> *Eriocaulon alaeum* Kimpouni in Fragm. Flor. Geobot. **39**: 326 (1994). Type: Congo, Katanga, Marungu Plateau, near Luonde, 14.vi.1969, *Lisowski, Malaisse & Symoens* 6534 (BR holotype, BRVU).

Slender rosulate annual. Leaves delicate, narrowly subulate to filiform, to 10 cm long, 0.3–0.6 mm wide, extended into a long bristle-like tip, yellowish-green. Scapes up to 20, slightly flexuous, 7–25 cm high, 4–6-ribbed; sheaths inflated, usually a little shorter than the leaves, the mouth deeply slit with acute limb. Capitulum hemispherical, 3–7 mm wide, pale buff or grey, the bracts sometimes edged in black or occasionally mostly blackish; involucral bracts cup-shaped in 2–3 series, obovate with broadly rounded tip, cartilaginous; floral bracts oblong-cuneate, scarious, glabrous or with a few inconspicuous papillae on back, tip subacute to apiculate, upper margin denticulate; receptacle pilose. Flowers trimerous, 1.3–2.3 mm long, subsessile, pale or the perianth parts with dark tips. Male flowers: sepals ligulate, unequally and variably joined, the lateral lightly keeled, ± glabrous, tips truncate-denticulate, emarginate; petals rudimentary at summit of stipe which equals calyx, glabrous, eglandular or laterals minutely glandular; anthers black. Female flowers: sepals resembling male, the two lateral falcate, spongy with large cells on the upper keel or sometimes broadened into a slight wing, keel sometimes with a few inconspicuous papillae, third sepal straight; petals unequal, spongy, all eglandular or only the median, ± glabrous, bidentate, two narrowly lanceolate-oblong, the median lanceolate; ovary raised on a stipe. Seeds plumply ellipsoid, 0.4 mm long, reddish-brown, closely transversely white-papillose.

Botswana. N: Moremi Game Reserve, off Mboma–Gadikwe channel, 19°11.4'S 23°16'E, 15.viii.1979, *P.A. Smith* 2802 (SRGH). **Zambia**. N: Mansa Dist., Kali, between Mansa and Lake Bangweulu, 17.ix.1911, *R.E. Fries* 634a (K, UPS). W: Shangombo Dist., Lui R., Namaenda, 10.viii.2006, *Bingham & Vestergaard* 13107 (K). **Malawi**. N: Nkhotakota Dist., between 1.6 km S of Dwambadgi and lakeshore, 13.vii.1950, *Foster* 7 (K).

Also in S Tanzania and S Congo (Katanga). Shallow water in boggy grassland; 500–1300 m.

Conservation notes: Widespread; not threatened.

E. stenophyllum can be recognised by its thin subfiliform leaves, glabrous capitula with pale, crateriform, cartilaginous involucral bracts, and by its narrow sepals. The amount of blackish coloration in the capitulum appears to be very variable and of no taxonomic significance.

E. bicolor Kimpouni, a dimerous offshoot from *E. stenophyllum* and known only from the type collection from Congo (Katanga), may occur in N Zambia. The female sepals have a wider wing than in *E. stenophyllum*.

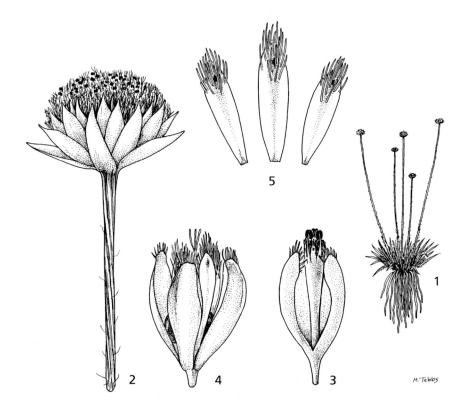

Fig. 13.4.6. ERIOCAULON MBALENSIS. 1, habit (× ²/₃); 2, capitulum (× 10); 3, male flower (× 20); 4, female flower (× 20); 5, petals of female flower (× 20), all from *Richards* 5967a. Drawn by Margaret Tebbs. From Kew Bulletin.

31. **Eriocaulon mbalensis** S.M. Phillips in Kew Bull. **52**: 71, fig.6f–k (1997); in Kirkia
    17: 44 (1998). Type: Zambia, Mbala Dist., Kambole, 25.viii.1956, *Richards* 5967a
    (K holotype). FIGURE 13.4.**6**.

Small perennial forming clusters of rosettes, the fine woolly hairs in the rosette centre
extending on to leaves and sheaths. Leaves narrowly lanceolate, tapering from base to acute tip,
1–2 cm long, 1.5–2 mm wide. Scapes few, 3–5 cm high, relatively stout, 5-ribbed, glabrous near
the capitulum, scattered fine silky hairs below the middle; sheaths about as long as the leaves, the
limb splitting into 2–3 teeth. Capitulum hemispherical, 5–5.5 mm wide, dull brown with shaggy
white-haired flowers intermingling with bract tips; involucral bracts as wide as the capitulum,
spreading at maturity, firm, dull brown with a paler central stripe, the outer whorl narrowly
lanceolate, acute, 3 mm long, the inner broader, ovate, sharply acute to cuspidate; floral bracts
resembling the inner involucral but more angular, cuspidate, glabrous; receptacle villous.
Flowers trimerous, 2 mm long. Male flowers: sepals almost completely joined into a spathe, the
upper margin shallowly 3-lobed, white-villous; petals shortly exserted, glandular, white-villous;
anthers black. Female flowers: sepals subequal, dull brown, oblanceolate-oblong, lightly concave,
not keeled, laterals slightly falcate, median straight, white-pilose along margin of the broadly
obtuse tip, the hairs forming an apical tuft; petals raised on a short stipe, subequal, narrowly
oblanceolate-oblong, glandular, white-villous towards the tip, a few slender translucent hairs
below. Seeds subglobose, 0.5 mm long, brown with longitudinal striations.

Zambia. N: Mpulungu Dist., Kambole, 1500 m, 25.viii.1956, *Richards* 5967a (K).
Known only from the type, which was collected in deep mud in a gathering of
Syngonanthus angolensis; c.1500 m.
Conservation notes: Known only from the type hence Data Deficient; probably
Vulnerable.

32. **Eriocaulon angustibracteum** Kimpouni in Fragm. Flor. Geobot. **39**: 329 (1994).
    —Phillips in F.T.E.A., Eriocaulaceae: 28 (1997); in Kirkia **17**: 45 (1998). Type:
    Congo, Katanga, Kisenge, *Duvigneaud* 3259 (BRLU holotype).

Rosulate annual. Leaves narrowly linear to linear-subulate, up to 10 cm long, 1–3 mm wide,
thin, septate, tapering to an acuminate, sometimes bristly extended tip. Scapes up to 20, 7–28 cm
high, 5–7-ribbed; sheaths shorter than the leaves, obliquely slit with an obtuse, often splitting limb.
Capitulum globose, 4–6 mm wide, blackish when young with paler involucral bracts, maturing
grey and buff, the white-pilose flowers visible among the bracts; involucral bracts obovate to
rotund with rounded tip, shorter than the capitulum width, pale straw or greyish tinged, firmly
scarious, reflexed at maturity; floral bracts narrowly cuneate or oblanceolate, grey-buff, glabrous
or the inner thinly white-pilose, acute to shortly acuminate; receptacle pilose. Flowers trimerous,
1.5–2 mm long. Male flowers: sepals oblong, the lateral keeled, almost free or joined below the
middle into a spathe, tips obtuse, white-pilose; petals small, subequal, eglandular, white-pilose;
stamens 3, epipetalous, black. Female flowers: sepals very unequal, the two lateral boat-shaped,
narrowed to the base, a fleshy wing on the centre keel, grey-buff or sepal body grey with paler
wing, wing width variable, when broad the margin often toothed, white-pilose on the back above
the wing, tip acute to shortly acuminate; median sepal shorter, linear-oblanceolate, the tip sparsely
white-pilose; petals narrowly linear, the median a little larger, eglandular, tips white-pilose. Seeds
ellipsoid, 0.4 mm long, brown, transversely white papillose-striate.

Zambia. B: Senanga Dist, Namushakende, 2.v.1964, *Verboom* 1732 (K, SRGH). N:
Mwense Dist, Chipili, 6.vi.1957, *Robinson* 2241 (K, LISC, SRGH). W: Ndola Dist.,
Mwekera near Kitwe, 24.iv.1956, *Mortimer* 143 (SRGH).
Also in S Congo (Katanga), Cameroon and S Tanzania. Marshy grassland and rice
fields, the leaf rosette often submerged; 1000–1400 m.
Conservation notes: Moderately widely distributed; not threatened.
E. angustibracteum is the only species of *Eriocaulon* in the Flora area to possess just
3 stamens.

33. **Eriocaulon deightonii** Meikle in Kew Bull. **22**: 143 (1968); in F.W.T.A., ed.2 **3**: 63 (1968). —Phillips in Kirkia **17**: 46 (1998). Type: Sierra Leone, near Modno, Mambolo, 15.x.1953, *Jordan* 946 (K holotype).

Eriocaulon acutifolium S.M. Phillips in Kew Bull. **52**: 59 (1996). Type: Zambia, Solwezi, 10.iv.1960, *Robinson* 3506 (K holotype).

Tufted annual, occasionally developing a short stem to 2 cm long. Leaves subulate, tapering to a finely acuminate tip, 4–13 cm long, 1.5–2.5 mm wide, thin, septate. Scapes many, up to about 50, 10–25 cm high, 5–6-ribbed; sheaths shorter than the leaves, the limb splitting into 2–3 short teeth. Capitulum subglobose, 4–4.5 mm wide, greyish-buff varying to grey with paler involucral bracts, glabrous or with a thin scattering of white hairs, slightly glossy; involucral bracts obovate-oblong to suborbicular with rounded tip, as wide as the capitulum, yellowish-grey; floral bracts obovate-spathulate, concave, incurving, acute, glabrous or thinly white-pilose on the upper back; receptacle hemispherical, pilose. Flowers trimerous, 1–1.6 mm. Male flowers: sepals joined into an open spathe with incurving margins, tips free, obtuse-denticulate, sparsely white-hairy; petals small, included within the calyx, subequal, glands small or absent, tips sparsely white-pilose; anthers black. Female flowers: sepals unequal, two lateral boat-shaped, broadly winged on the centre keel, narrowed to the base, acute, a few small white papillae on the keel and margins, median sepal narrowly elliptic, as long as the laterals but narrower, scarcely winged, petals subequal, linear-oblanceolate, eglandular, glabrous or a few white hairs near tip. Seeds ellipsoid, 0.4 mm long, light brown, transversely white papillose-striate.

Zambia. W: Solwezi, 1350 m, 10.iv.1960, *Robinson* 3506 (K).

Also in Guinea, Sierra Leone and Ivory Coast; boggy grassland, swamp margins and damp hollows, the leaf rosette often submerged; 1350 m.

Conservation notes: Widely distributed, although very localised within the Flora area. Data Deficient in the Flora area, but probably not threatened globally.

E. deightonii resembles *E. buchananii* in floral and seed morphology, but is easily distinguished by its much longer, fine-tipped leaves. *E. angustibracteum* also has similar flowers and long leaves with fine tips, but this usually has somewhat larger capitula (4–6 mm wide) and only three stamens.

34. **Eriocaulon buchananii** Ruhland in Bot. Jahrb. Syst. **27**: 83 (1899). —Hess in Ber. Schweiz. Bot. Ges. **65**: 145 (1955). —Kimpouni in Fragm. Flor. Geobot. **39**: 331 (1994). —Phillips in F.T.E.A., Eriocaulaceae: 28, fig.3 (1997); in Kirkia **17**: 46 (1998). Types: Malawi, no locality, 1891, *Buchanan* 1168 (B† syntype, BM, K) & Tanganyika Plateau, Chitipa (Fort Hill), vii.1896, *Whyte* s.n. (B† syntype, BM, K, P, Z). FIGURE 13.4.4.

Slender tufted annual. Leaves in a small basal rosette, pale bronze-green, broadly linear, 1–3 cm long, 1–3 mm wide, spongy, subacute. Scapes up to 50 or more in an erect bunch, slender, 7–20 cm high, 4–5-ribbed; sheaths equalling the leaves, the limb scarious, acute. Capitulum globose to slightly ovoid, greyish-brown or blackish, 4–5 mm wide, hard, glossy, completely glabrous, floral bracts loosely imbricate, the flowers scarcely visible among them; involucral bracts small, shorter than the capitulum width, light brown, firm, obovate-oblong, the tips rounded and incurved, reflexed at maturity; floral bracts angular-spathulate, ± flat, grey, glabrous, acute; receptacle villous. Flowers trimerous, 1.2–1.5 mm long, pedicellate. Male flowers: calyx spathate, thin, brownish-grey, the free lobes subtruncate; petals very small, unequal, eglandular, the longest exserted from the calyx with a few small white papillae at the tip; anthers black. Female flowers: sepals very unequal, the two lateral deeply concave, frequently obtriangular, the keel gibbous with a broad spongy wing, narrowed to the base, tip acute, a few translucent hairs often present inside, otherwise glabrous, third sepal linear to narrowly oblong; petals slightly unequal, narrowly oblanceolate-oblong, eglandular or the two smaller with small glands, sometimes emarginate, longest exserted from the calyx and often tipped with a few small white papillae. Seeds ellipsoid, 0.35 mm long, light brown to reddish-brown, transversely white papillose-striate.

Zambia. N: Mansa Dist., 45 km N of Mansa (Fort Roseberry), 5.vi.1960, *Robinson* 3733 in part (mixed with *E. mutatum*) (SRGH). W: Kitwe, 13.vi.1963, *Mutimushi* 317 (K, SRGH). C: Kabwe (Broken Hill), v.1909, *Rogers* 8104 (K, SRGH). E: Katete Dist, Katete Hills, mushitu, 10.vi.1963, *Verboom* 140 (K). S: Namwala Dist, 55 km NW of Kasenga, 7.vii.1963, *Robinson* 5541 (K). **Zimbabwe**. N: Darwin Dist., Msengesi camp, 13.v.1955 *Whellan* 935 (K, SRGH). W: Matobo Dist., Besna Kobila farm, v.1959, *Miller* 5922 (K, LISC, SRGH). C: Harare Dist., Hatfield, 8.v.1934, *Gilliland* 85 (K, SRGH). **Malawi**. N: Chitipa (Fort Hill), vii.1896, *Whyte* s.n. (K). S: Chiradzulu Dist., Namadsi R., 19.xii.1899, *Cameron* 50 (K).

Also in Burundi, Tanzania and Angola. Shallow water in pans, and wet sand or mud in recently flooded, drying hollows and in boggy grassland; 1000–1400 m.

Conservation notes: Widely distributed, not threatened.

The bunch of many erect, slender scapes topped by hard, globose capitula is characteristic of *E. buchananii*. The young heads are black (except for the paler involucral bracts), but frequently lighten to a greyish-brown at maturity. The spongy gibbous 'shoulders' of the female sepals are often visible among the loosely arranged floral bracts.

It is closely related to the West African *E. fulvum* N.E. Br., which is mainly distinguished by its pale capitula and obtuse floral bracts. *E. maculatum* and *E. strictum* (from Tanzania) also belong to this complex of erect annuals with very unequal female sepals and similar seeds, mostly with rows of white papillae. The boundaries between the individual species are not clear cut, and intermediate specimens may be encountered which are not easily assigned. *Robinson* 2852 (Zambia S:) has pale capitula and female sepals with a translucent patch as in *E. fulvum*, *E. maculatum* and *E. strictum*, but otherwise resembles *E. buchananii*.

E. buchananii has very similar seeds to *E. transvaalicum*, another annual with many scapes of globose blackish capitula, but the latter has larger leaves and capitula, stiff, straight, diverging scapes and lacks the gibbous and spongily-winged female sepals so characteristic of *E. buchananii*.

35. **Eriocaulon chloanthe** S.M. Phillips in Kew Bull. **51**: 637 (1996); in Kirkia **17**: 48 (1998). Type: Zambia, South Luangwa Nat. Park, 10.v.1963, *Verboom* 132 S (K holotype). FIGURE 13.4.**7**.

Rosulate herb, probably annual. Leaves absent at maturity. Scapes c.20, stout, stiff, 18–23 cm high, 5–ribbed. Capitulum globose, hard, 5 mm wide, pale grey-brown, slightly glossy, completely glabrous, bracts neatly imbricate, the flowers hidden except for the spongy sepal-wings at maturity; involucral bracts small, concolorous with the floral, obovate-oblong, firm, reflexed at maturity; floral bracts pale brown with greyish tips, firmly scarious, narrowly obovate, acute; receptacle villous. Flowers trimerous, 1.2–1.6 mm long, pedicellate. Male flowers delicate, pallid flushed pale grey: sepals completely joined into a funnel-shaped cup open only at the top, margin erose; petals small, included within the calyx, eglandular, glabrous or the largest with one or two white papillae; anthers black. Female flowers: sepals very unequal, the laterals deeply concave, strongly gibbous with a broad, spongy, swollen wing on the keel, the margins flaring outwards, a few long translucent hairs towards the base inside, a thin dark patch of translucent tissue on the body between wing and margins, subacute; median sepal shorter, linear; petals raised on a short stipe 0.2 mm long, linear-oblanceolate, eglandular, glabrous, the median a little larger. Seeds ellipsoid, 0.4 mm long, brown, transversely white-reticulate.

Zambia. E: Chipata Dist., South Luangwa Nat. Park (Nsefu Game Res.), 10.v.1963, *Verboom* 132 S (K).

Known only from the type collection, which lacks habitat details; c.600 m.

Conservation notes: Known only from the type; Data Deficient, but probably Vulnerable.

The only specimen lacks leaves, but its combination of floral and seed morphology clearly excludes it from other species in the group. The funnel-shaped non-spatheate male calyx, open at the top, is unique in this group.

36. **Eriocaulon mulanjeanum** S.M. Phillips in Kew Bull. **51**: 638 (1996); in Kirkia **17**: 48 (1998). Type: Malawi, Mt Mulanje, Lichenya Plateau, 8.v.1963, *Wild* 6162 (K holotype, LISC). FIGURE 13.4.**7**.

Small erect annual. Leaves few, basal, linear, 1–2 cm long, c.1 mm wide, thin, septate, acute. Scapes 1–4, rather stiff, 3–5 cm high, 4-ribbed; sheaths equalling the leaves, slightly inflated. Capitulum globose, black, 3.5–4 mm wide, glabrous, loosely few-flowered; involucral bracts in a single series, some with flowers in the axils, obovate to obovate-oblong, black, scarious, concave with rounded tip, smaller than the capitulum width at maturity; floral bracts resembling the involucral in colour and texture, narrowly obovate-oblong, concave and rounded; receptacle pilose. Flowers black, trimerous, completely glabrous, subsessile. Male flowers: 1.2 mm long, sepals joined into a funnel-shaped spathe split halfway on one side, truncate-erose across the top; petals tiny, eglandular; anthers black. Female flowers 1.6–2 mm long: sepals very unequal, the laterals deeply concave and strongly gibbous, a prominent, crescent-shaped, swollen, spongy wing along the midline, sepal margins also spongy, wing-like and flaring outwards as a circular frill, sepal body thin and translucent, tip rounded; median sepal lanceolate, concave, unwinged; petals narrowly oblanceolate-oblong, eglandular, obtuse, the median somewhat larger. Seeds ellipsoid, 0.5 mm long, brown, reticulate with white papillae in irregular longitudinal lines.

Malawi. S: Zomba Dist., Zomba Mt, near Chivunde Forestry village, 8.iv.1971, *Jordan* 3001 (K).

Only known from two collections on mountains in S Malawi. Forming colonies among moss in wet patches; 1600–1800 m.

Conservation notes: Endemic to southern Malawi, very localised but within protected areas; probably Vulnerable.

In most *Eriocaulon* species the most obvious feature of the capitulum is the overlapping floral bracts, the flowers being usually hidden or at most some petals exserted. However, in *E. mulanjeanum* the protruding swollen, crescent-shaped female sepals are more obvious than the intermingling bracts. It closely resembles *E. maculatum* in floral structure, but this has a brownish capitulum with regularly imbricated, subacute floral bracts which ± obscure the flowers.

37. **Eriocaulon maculatum** Schinz in Bull. Herb. Boiss., sér.2 **6**: 709 (1906). — Obermeyer in F.S.A. **4**(2): 14 (1985). —Phillips in F.T.E.A., Eriocaulaceae: 31 (1997); in Kirkia **17**: 49 (1998). Type: South Africa, Limpopo Prov., Blouberg, 10.iii.1894, *Schlechter* 4651 (Z holotype, K, PRE).

Small rosulate annual. Leaves linear, 0.5–1.5 cm long, 1–2 mm wide, apiculate. Scapes up to 20, straight, 3–12 cm high, 4-ribbed; sheaths equalling the leaves, the mouth 2–3-fid. Capitulum globose to slightly dome-shaped, 3–4 mm wide, brown-grey, floral bracts loose with the darker flowers visible between; involucral bracts as wide as the capitulum, pinkish-brown with paler margins, scarious, obovate with rounded tips, reflexed at maturity; floral bracts resembling the involucral in colour and texture, oblanceolate to spathulate, glabrous, subacute; receptacle cylindrical, villous. Flowers trimerous, 1–1.5 mm long, stipitate. Male flowers: calyx funnel-shaped, spathate with a truncate-erose margin, glabrous; petals included within the calyx, tiny, glabrous and eglandular; anthers black. Female flowers obtriangular; sepals very unequal, two deeply boat-shaped, the keel gibbous with a pale, thick, spongy wing surrounding a thinner blackish central patch, margins flaring outwards with a few translucent hairs inside, median sepal reduced, linear, sometimes as long as the other two but usually much shorter, sometimes vestigial; petals linear-oblanceolate, glabrous and eglandular, the median slightly larger. Seeds ellipsoid, 0.35 mm long, reddish-brown with a white reticulate patterning, glossy.

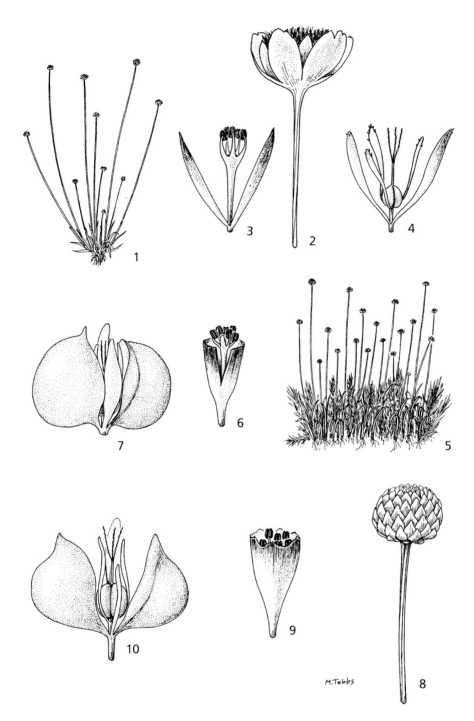

Fig. 13.4.**7**. ERIOCAULON FUSCUM. 1, habit (× ²/₃); 2, capitulum (× 10); 3, male flower (× 30); 4, female flower (× 30). E. MULANJEANUM. 5, habit, drawn among moss (× ²/₃); 6, male flower (× 20); 7, female flower (× 20). E. CHLOANTHE. 8, capitulum (× 4); 9, male flower (× 20); 10, female flower (× 20), 1–4 from *Gonde* 210, 5–7 from *Wild* 6162, 8–10 from *Verboom* 132S. Drawn by Margaret Tebbs. From Kew Bulletin.

Zimbabwe. C: Makoni Dist, near Rusape, 1922–1926, *Hislop* 413 (K). **Malawi**. S: Mulanje Dist., 16 km NW of Likabula, 15.vi.1962, *Robinson* 5358A (SRGH). **Mozambique**. MS: Muanza Dist., Cheringoma sawmill, Chiniziua R., 13.vii.1972, *Ward* 7893 (SRGH).

Also in N South Africa and S Tanzania. Reported from Zambia by Obermeyer (1985). Boggy grassland and margins of drying pools on sandy soils; 100–1000 m.

Conservation notes: Widespread distribution, not threatened.

E. strictum Milne-Redh. from Mafia Is.,Tanzania, is a very similar small annual. It differs from *E. maculatum* in its narrower, almost filiform, smooth leaves and by the absence of the median sepal in the female flowers. The seed-coat patterning is also different.

2. **MESANTHEMUM** Körn.

Mesanthemum Körn. in Linnaea **27**: 572 (1856). —Ruhland in Engler, Pflanzenr. **13**: 117–119 (1903). —Jacques-Felix in Bull. Soc. Bot. France **94**: 143–151 (1947). —Hess in Ber. Schweiz. Bot. Ges. **65**: 178–185 (1955). —Kimpouni in Fragm. Flor. Geobot. **39**: 147–160 (1994). —Phillips in Kirkia **17**: 49–55 (1998).

Perennials from a stout tough rhizome, rarely slender annuals (annuals confined to West Africa), leaves and scapes frequently hairy, old leaves glabrescent; leaves in a basal rosette, linear, spongy. Scapes unbranched, many-ribbed, arising from the leaf axils; sheaths obliquely slit, the limb acute. Capitulum globose or flattened; involucral bracts in several imbricate series, coriaceous with scarious margins, innermost sometimes radiating beyond the periphery of the white-hairy floral disc; floral bracts and flowers embedded in a woolly cushion of long receptacular hairs; floral bracts filiform with expanded hairy tips; flowers trimerous, pedicellate. Male flowers: sepals usually membranous, oblanceolate-oblong, concave, free or basally joined; petals united into a fleshy funnel-shaped or oblong tube with a basal stipe and shallowly 3-lobed upper margin, petal bases sometimes free resulting in slits in the petal-tube, 3 epipetalous glands within; stamens 6, arising from base of petal-tube; anthers yellowish, exserted above the corolla rim at anthesis; vestigial gynoecium present. Female flowers: sepals free, resembling the male sepals, soon falling; petals free at base around the ovary, joined above into a fleshy cylindrical tube, this equalling or longer than the sepals, glabrous or villous both outside and within, 3 linear brownish glands inserted about $^2/_3$ up inside, densely white-pilose at tip. Seeds subglobose, brown, densely covered in white hair-like projections.

15 species in tropical Africa, 2 endemic in Madagascar.

The only widespread species is *M. radicans*. The remainder are mostly of local distribution, concentrated especially in West Africa and from S Congo to N Zambia. Most species are robust rhizomatous perennials, unlikely to be confused with any but the most vigorous perennial species of *Eriocaulon*. The coriaceous involucral bracts are also tougher than is usual in *Eriocaulon*.

1. Capitulum flat-topped at maturity; involucre bowl-shaped or obconical, the bracts clearly visible; sepals pallid · 2
– Capitulum subglobose at maturity; involucre flat at first, the bracts later reflexing and obscured; sepals often brown or blackish · 4
2. Innermost involucral bracts extending beyond floral disc; leaves 9–14 mm wide; petals of female flowers villous on both faces · · · · · · · · · · · · · · · · · **1**. *radicans*
– Innermost involucral bracts equalling floral disc; leaves 3–8 mm wide · · · · · · 3
3. Involucral bracts broadly ovate without a conspicuous margin, appressed-silky, nerves obscure; hairs on receptacle wiry, blackish; petals of female flowers villous on outer face, glabrous within · **2**. *variabile*

- Involucral bracts ovate-oblong with conspicuous paler margins, glabrous, nerves prominent; hairs on receptacle soft, grey-brown; petals of female flowers villous on both faces · **3.** *glabrum*
4. Involucral bracts 4.5–6 mm long; female flowers with circlet of long hairs from pedicel-tip and base of sepals; petals of female flowers villous on outer face only · **4.** *pilosum*
- Involucral bracts c.3 mm long; female flowers with glabrous pedicel-tip and sepal-bases; petals of female flowers villous on both faces· · · · · · · · **5.** *africanum*

1. **Mesanthemum radicans** (Benth.) Körn. in Linnaea **27**: 573 (1856). —N.E. Brown in F.T.A. **8**: 260 (1901). —Ruhland in Engler, Pflanzenr. **13**: 119 (1903). —Hess in Ber. Schweiz. Bot. Ges. **65**: 180 (1955). —Kimpouni in Fragm. Flor. Geobot. **37**: 134 (1992). —Phillips in F.T.E.A., Eriocaulaceae: 34 (1997); in Kirkia **17**: 50 (1998). Types: Sierra Leone, no locality, *Don* s.n.; Liberia, Grand Bassa, *Ansell* s.n.; and Angola, no locality, *Curror* s.n. (all K syntypes). FIGURE 13.4.8.

   Eriocaulon radicans Benth. in Hooker, Niger Fl.: 547 (1849).

   Mesanthemum erici-rosenii T.C.E. Fr. in R.E. Fries, Wiss. Ergebn. Schwed. Rhod.-Kongo-Exped. **1**: 218 (1916). —Kimpouni in Fragm. Flor. Geobot. **37**: 134 (1992). Type: Zambia, Lake Bangweulu (Bangweolo), Mbawala Is., ix.1911, *von Rosen* 806 (UPS holotype).

Robust tussocky perennial from a short or elongating rhizome. Leaves clustered at the rhizome tip, 12–50 cm long, (5)10–15 mm wide, scattered-pilose or glabrescent. Scapes 30–60 cm high; sheaths usually pilose. Capitulum creamy-white, flat-topped with a bowl-shaped involucre, 10–15 mm wide; involucral bracts straw-coloured with greenish conspicuously-nerved tips and broad pale margins, patchily appressed silky-hairy, outermost 3–3.6 mm long, broadly ovate, rounded, innermost extending shortly beyond the floral disc, oblong, obtuse or subacute, appressed white-villous on the inner face; floral bracts capillary with a white-pilose subulate tip; receptacular hairs dark grey. Flowers 2.2–2.8 mm long, sepals pallid. Male flowers: sepals free, broadly oblong, tips truncate and white-ciliate. Female flowers: sepals ovate-oblong, truncate-denticulate, glabrous or the tips thinly ciliate; petal-tube villous with long grey hairs outside and within, tips densely pilose with short white hairs.

Zambia. N: Samyfa Dist., Lake Bangweulu, Mbawala Is., ix.1911, *von Rosen* 806 (UPS). **Mozambique**. Z: Maganja da Costa Dist., 14 km from Vila da Maganja to Malri, 50 m, 21.xi.1967, *Torre & Correia* 16178 (LISC). MS: Cheringoma coast, Nyamaruza dambo, v.1973, *Tinley* 2915 (K, LISC, SRGH).

Also in East Africa, Angola, Congo, Congo Republic, Gabon and West Africa. Swampy grassland, lake margins and acid sand near the coast; 5–1200 m.

Conservation notes: Widely distributed; not threatened.

This is the most widespread species of *Mesanthemum* in Africa, but apparently is only common in West Africa. It is known in East Africa from the vicinity of Lake Victoria, Lake Bangweulu, and from coastal Tanzania and Mozambique.

M. radicans resembles *M. variabile* in its ovate, appressed silky-hairy involucral bracts, but differs in the broad pale margins and the conspicuously nerved tips of these bracts. The long, narrow innermost involucral bracts are white-villous on the exposed inner face. On dissection, *M. radicans* can also be distinguished from *M. variabile* by the dense, long grey hairs inside the petal-tube of the female flowers.

M. cupricola Kimpouni from Congo (Katanga) also has inner involucral bracts which exceed the floral disc, but differs from *M. radicans* by its acute outer bracts and by the petals of the female flowers which are glabrous (not villous) on the inner face.

Fig. 13.4.8. MESANTHEMUM RADICANS. 1, habit (× ⅔); 2, capitulum (× 2); 3, outer involucral bract (× 4); 4, inner involucral bract (× 4); 5, floral bract (× 14); 6, male flower (× 14); 7, inside view of male petal-tube with stamens (× 20); 8, female flower (× 14); 9, inside view of female petal-tube (× 20), all from *Norman* 54. Drawn by Margaret Tebbs. From Flora of Tropical East Africa.

2. **Mesanthemum variabile** Kimpouni in Fragm. Flor. Geobot. **39**: 157 (1994). —
   Phillips in Kirkia **17**: 51 (1998). Type: Congo, Makia Plateau, 3 km W of Katema,
   19.i.1969, *Lisowski, Malaisse & Symoens* 423 (POZG holotype, BR, BRVU).

Tufted perennial from a short rhizome. Leaves narrowly linear, 13–30 cm long, 4–8 mm
wide, pubescent to hirsute. Scapes 40–70 cm high, pubescent; sheaths with long acuminate
limb. Capitulum flat-topped with a bowl-shaped involucre, 13–15 mm wide, involucre buff,
floral disc white, innermost bracts equalling disc; involucral bracts firmly cartilaginous,
broadly ovate, nerves obscure, the outer subacute, 2.5–4 mm long, inner sometimes obtuse, all
appressed silky-hairy on the back, the innermost sometimes also on the inner face; floral
bracts capillary with an obovoid white-pilose tip; receptacular hairs blackish, coarse, wiry.
Flowers 2.7–3 mm long; sepals pallid. Male flowers: sepals free, oblanceolate-oblong, concave,
white-pilose on the upper back and rounded tips. Female flowers: sepals obovate-oblong to
narrowly oblong, tips truncate-dentate with a few white hairs on the teeth; petals clawed and
free in the lower third, united into a spongy tube above, white-pilose towards the subtruncate
tip or from the middle upwards, villous with long, dark grey hairs arising near the margins at
top of the free claws outside, a few marginal hairs sometimes also projecting inside the petal-
tube, otherwise glabrous within.

Zambia. N: Kawambwa Dist., Ntumbachusi (M'tunatusha) R., 28.xi.1961, *Richards*
15412 (BR, K). W: Mwinilunga Dist., Zambezi rapids, 6 km N of Kalene Hill,
21.ii.1975, *Hooper & Townsend* 296 (K).

Also in S Congo. Wet hollows and marshy ground near rivers; 1200–1600 m.
Conservation notes: Restricted distribution; probably Near Threatened.

M. variabile has silky involucral bracts like *M. radicans*, but is a more slender species
with narrower, hairier leaves and an involucre not extending beyond the floral disc.
It can also be recognized by its coarse black receptacular hairs and blackish sepals,
and the interior to the female petal-tube is glabrous.

3. **Mesanthemum glabrum** Kimpouni in Fragm. Flor. Geobot. **39**: 153 (1994). —
   Phillips in Kirkia **17**: 53 (1998). Type: Congo, Kasai, Kapanga, x.1933, *Overlaet*
   807 (BR holotype).

Rosulate perennial. Leaves linear or subulate, up to 40 cm long, 3–6 mm wide, glabrous on
the lower surface, hirsute above or glabrescent, tip thickened, rounded. Scapes 40–75 cm high,
glabrous to hirsute; sheaths hirsute, the limb narrowed to a shortly rostrate obtuse tip.
Capitulum 9–14 mm wide, flat topped with an obconical involucre, floral disc densely white-
pilose, inner involucral bracts not exceeding the floral disc; involucral bracts brownish-green
with a broad pale scarious margin, strongly ribbed in upper half, smooth below, tips obtuse to
broadly rounded, essentially glabrous (some short marginal cilia), outermost ovate-oblong,
3.7–5 mm long, the inner progressively longer and more oblong, innermost 4.5–5.5 mm long;
floral bracts filiform, a dense brush of white hairs at the expanded tip; receptacular hairs pale
grey-brown. Flowers 3–4 mm long; sepals pallid. Male flowers: sepals joined at the extreme base,
tips white-pilose. Female flowers: sepals resembling the male, glabrous on the back, white-pilose
at the truncate denticulate tips; petals narrowly oblong, variably free up to about halfway, joined
above, villous with long grey-brown hairs both outside and within up to level of glands, tips
densely white-pilose.

Zambia. N: Chinsali Dist., Ishiba Ngandu (Shiwa Ngandu), Lake Young, 15.i.1959,
Richards 10656 (K); Chama Dist., Manshya R., 28.xii.1963, *Symoens* 10781 (BR, BRLU,
K). W: Mwinilunga Dist., Dobeka dambo, 54 km W of Mwinilunga, 22.i.1975,
Brummitt, Chisumpa & Polhill 13998 (K).

Also in Angola and S Congo. Wet soil in dambos; 1200–1500 m.
Conservation notes: Fairly widely distributed; not threatened.

A relatively slender species with an involucre of neat, rather narrow, conspicuously
pale-margined and prominently nerved glabrous bracts.

Drummond & Cookson 6444 (Zambia B: Kalabo, 13.xi.1959) resembles *M. glabrum*, but has broader, dark brown involucral bracts, fleshier glabrous leaves, and a brush of hairs from the pedicel-tip of the male flowers. It may prove to be an undescribed species when more material is available.

4. **Mesanthemum pilosum** Kimpouni in Fragm. Flor. Geobot. **39**: 155 (1994). — Phillips in Kirkia **17**: 54 (1998). Type: Congo, Kundelunga Plateau, 27.x.1969, *Lisowski, Malaisse & Symoens* 7438 (POZG holotype, BR, BRVU).

Robust tussocky perennial from a woody rootstock or short rhizome, old burned leaf remains often present at base; leaves, sheaths and scapes hairy. Leaves linear, 15–30 cm long, 8–12 mm wide, usually pilose on both surfaces but varying from glabrescent or pubescent to velvety; scapes 25–75 cm high. Capitulum hemispherical becoming subglobose at maturity, 12–16 mm wide, floral disc white and woolly, involucre brownish-green; involucral bracts equalling the floral disc, flat across capitulum base at first, reflexing at maturity, glabrous or puberulous with obvious nerves, margins ciliate, subacute, outer bracts ovate, 4.5–6 mm long, inner more oblong, white-tomentose on inner face towards tip; floral bracts filiform with a spathulate white-woolly tip; receptacular hairs grey. Flowers 2.5–4 mm long; sepals brownish-grey. Male flowers: sepals free, thinly cartilaginous, narrowly oblong, lightly keeled, tips white-pilose. Female flowers surrounded by a dense circlet of long, soft, grey hairs from the pedicel-tip and sepal-bases; sepals thin, half as long as the petal-tube, obovate-oblong, long-villous at base, hairy also on the back and margins, tips obtuse to dentate, white-pilose; petals unequally free to the middle or beyond, villous around the middle on the outer face with translucent hairs and glabrous within, joined and membranous above, tips white-pilose.

Zambia. B: Senanga Dist., 16 km N of Senanga, 31.vii.1952, *Codd* 7297 (K, PRE). N: Kawambwa Dist., Ntenke, 10.ix.1963, *Mutimushi* 454 (K). W: Mwinilunga Dist., SW of Dobeka bridge, 13.x.1937, *Milne-Redhead* 2748 (BR, K). C: Kabwe Dist., W side of Great North Road just beyond Mulunguishi R., 24 km N of Kabwe (Broken Hill), 23.ix.1947, *Brenan & Greenway* 7938 (K). **Malawi**. N: Mzimba-Nkhata Bay Dist., Vipya Plateau, 51 km SW of Mzuzu, 11.xi.1973, *Pawek* 7488 (K, SRGH).

Also in Congo (Katanga) and Angola. Boggy areas in grassland; 1000–1700 m.

Conservation notes: Fairly widely distributed; not threatened.

The long circlet of silky hairs surrounding the disarticulated female flower is a good spot character for this species.

5. **Mesanthemum africanum** Moldenke in Phytologia **3**: 113 (1949). —Phillips in Kirkia **17**: 54 (1998). Type: Mozambique, Chimanimani Mts, 9.vi.1948, *Munch* 72 (NY holotype, K).

Rosulate perennial from a woody rootstock. Leaves broadly linear, 10–35 cm long, 9–12 mm wide, softly and rather thinly pilose, older leaves glabrescent. Scapes 30–60 cm high, glabrous except below the capitulum; sheaths pilose. Capitulum hemispherical at first, later depressed-globose, densely white-hairy with intermingling black sepals, 10–14 mm wide, involucre slightly smaller than the capitulum width, light brown, reflexed and obscured at maturity; involucral bracts hirsute, ovate, acute or obtuse, 3.2 mm long, cartilaginous with a scarious margin, glossy brown on the inner face, inner bracts with an apical patch of white hairs inside; floral bracts filiform with a clavate, white-pilose tip; receptacular hairs black. Flowers 2–3 mm long; sepals black. Male flowers: sepals free, membranous, oblanceolate, white-pilose across the truncate tips; petals densely white-pilose at tips. Female flowers: sepals oblanceolate-oblong to obovate, tips erose-truncate, white-pilose, otherwise glabrous; petals free below the middle, villous with long black hairs on both faces, the hairs arising near petal bases, tips densely white-pilose.

Zimbabwe. E: Chimanimani Mts, xi.1947, *McCosh* 2 in SRGH 17692 (K, SRGH). **Mozambique**. MS: Chimanimani Mts, 8.vi.1949, *Munch* 213 (K, SRGH).

Endemic to the Chimanimani mountains; 1200–2400 m.

Conservation notes: Very restricted distribution, but not under any specific threat.

The long grey hairs on the petal-tube of the female flowers arise near the base of the petals, not spread over the central portion, as in the similar but more widespread *M. pilosum.*

3. **SYNGONANTHUS** Ruhland

Syngonanthus Ruhland in Urban, Symb. Antill. **1**: 487 (1900). —Phillips in Kew Bull. **52**: 73–89 (1997); in Kirkia **17**: 55–64 (1998).

Annuals or perennials, usually hairy, hairs frequently glandular. Leaves in a basal rosette, linear to subulate or needle-like, woolly in the axils. Scapes usually 3-ribbed; sheaths obliquely slit. Capitulum scarious, white or brown, pilose to almost glabrous; involucral bracts in several series, glabrous or softly ciliate on the margins, sometimes also on the back, the inner translucent and subtending flowers; floral bracts absent; receptacle woolly with long hairs surrounding the flowers. Flowers trimerous, pedicellate, pedicels villous. Male flowers: sepals lanceolate, lightly keeled, joined below the middle into a funnel-shaped tube with free lobes; petals completely united into a membranous, subtruncate tube; stamens 3, filaments joined to the inside of the corolla, extended above the corolla rim and bearing 3 white or yellowish anthers; a 3-branched rudimentary gynoecium at base of the corolla tube, branches with swollen glandular tips; after anthesis the filaments collapse inwards and the corolla tube closes over the anthers. Female flowers: sepals free, otherwise resembling the male; petals delicate, usually pilose, joined near the tips, the clawed bases free, the small free apical lobes incurled; ovary 3-carpellate, the carpels protruding between free petal-bases, style forming a hollow tube divided at the tip into 3 long stigmatic branches, usually with 3 alternating swollen-tipped glandular appendages which entangle with the incurled petal-tips after anthesis to form a clavate structure. Seeds plumply ellipsoid to cylindrical, usually longitudinally white-striped.

About 200 species, mostly in South America; 15 species known from tropical Africa and South Africa.

The above description applies only to the African species. In America the genus is much more diverse, sometimes with an elongate, branching stem and with a greater range of inflorescence and floral morphology. In Africa *Syngonanthus* is a critical genus of closely related species, mostly difficult to distinguish from one another. In contrast to *Eriocaulon*, flower structure is very uniform and specific delimitation rests mainly on small differences in vegetative and capitulum morphology. The scapes are frequently densely hairy below the capitulum, but appressed-hairy between the ribs or glabrescent lower down. Glandular-capitate hairs are always spreading.

It is not usually necessary to dissect the flowers further than an inspection of the sepals of the female flowers. The tubercle-based hairs on the sepals often fall off during dissection, but their position can be seen by the remaining tubercles, especially along the sepal margins. The presence or absence of swollen-tipped appendages alternating with the 3 stigmatic branches is important taxomonically, but is very difficult to see in the tiny flowers without good magnification. Only two African species lack these appendages, and will not often be found in the Flora area. It is only worth searching for this character if the specimen is a small annual with a capitulum not exceeding 4 mm wide.

1. Scapes 20–60 cm high or if less capitulum dark golden-brown; perennials from a rhizome or rootstock · 2
– Scapes up to 25 cm high; usually slender annuals; capitulum white or white with a yellowish-brown involucre· 4

2. Capitulum golden-brown; scapes many, less than 25(30) cm high; leaves 1–3 cm long, needle-shaped, recurved, in dense woolly-centred rosettes; sheath-limb spathe-like ·· **1.** *wahlbergii*
– Capitulum white or the involucre yellowish-brown; scapes 1–10, 20–60 cm high; leaves linear, up to 8 cm long; sheath-limb straight ···················· 3
3. Involucral bracts whitish, obtuse; sepals of female flowers pilose only on the margins; sheaths glandular-pilose with an abruptly acute limb; seed cylindrical, 0.7 mm long ··· **2.** *poggeanus*
– Involucral bracts brown, subacute; sepals of female flowers pilose on margins and back; sheaths softly pilose with an evenly tapering limb; seed ellipsoid, 0.5 mm long ·· **3.** *angolensis*
4. Capitulum 4–8 mm wide ·· 5
– Capitulum 1.5–4.5 mm wide (if more than 4 mm, a dwarf plant with scapes under 5 cm high) ··· 7
5. Scapes with a mixture of glandular-capitate and pointed hairs; involucral bracts lanceolate-oblong, yellowish brown ······················ **6.** *longibracteatus*
– Scapes with glandular-capitate hairs only; involucral bracts lanceolate, completely white or brown only near the scape ······················· 6
6. Involucral bracts yellow near the scape, white around the capitulum periphery, soft, subacute, mostly pilose on the back ···················· **4.** *robinsonii*
– Involucral bracts white, stiff, scarious, sharply acute, all except the innermost glabrous on the back ······························· **5.** *mwinilungensis*
7. Capitulum 4–4.5 mm wide; swollen-tipped appendages alternating with the 3 stigmas present ·· **7.** *exilis*
– Capitulum 1.5–3.8 mm wide; swollen-tipped appendages alternating with the 3 stigmas absent or incomplete ································ 8
8. Involucral bracts yellowish, ovate, slightly exceeding flowers; capitulum 3–3.8 mm wide; female flowers c.1 mm long, sepals acute ··········· **8.** *paleaceus*
– Involucral bracts white, lanceolate, clearly exceeding flowers; capitulum 1.5–2.5 mm wide; female flowers 0.5–0.7 mm long, sepals obtuse ······· **9.** *welwitschii*

1. **Syngonanthus wahlbergii** (Körn.) Ruhland in Engler, Pflanzenr. **13**: 247 (1903). —Hess in Ber. Schweiz. Bot. Ges. **65**: 186 (1955). —Obermeyer in F.S.A. **4**(2): 19 (1985). —Phillips in F.T.E.A., Eriocaulaceae: 35 (1997); in Kirkia **17**: 56 (1998). Type: South Africa, former Transvaal?, *Wahlberg* s.n. (S holotype). FIGURE 13.4.**9**.

   Paepalanthus wahlbergii Körn. in Martius, Fl. Bras. **3**: 459 (1863). —N.E. Brown in Fl. Cap. **7**: 59 (1897); in F.T.A. **8**: 263 (1901).

   Eriocaulon recurvifolium C.H. Wright in Bull. Misc. Inform., Kew **1919**: 264 (1919). Type: Congo, Atené, i.1914, *Vanderyst* 3133 in part (K holotype) (mixed with *S. schlechteri* Ruhland).

Small tussocky perennial, the leaf-rosettes conspicuously woolly in centre, growing in clusters from a branching rhizome. Leaves numerous, needle-like, 1–3 cm long, 0.4–0.8 mm wide, thinly appressed-pilose to subglabrous, the narrowly obtuse tips often curving inwards. Scapes several to many flowering in succession, less than 25(30) cm high, 3-ribbed, pilose with patent glandular hairs especially immediately below the capitulum, some appressed pointed hairs also present; sheaths glandular-hairy, loose, the mouth inflated with open spathe-like limb. Capitulum golden or dark brown, 5–6 mm wide, bracts and sepals concolorous with the white petal-tubes visible between; outer involucral bracts oblong, obtuse, 1.5–1.8 × 0.8–1 mm, the inner progressively longer and narrowly elliptic-oblong, glabrous, innermost 2–2.5 mm long, subacute, ciliate on margins; receptacle villous. Flowers 1.5 mm long, brown; sepals of the female flowers lanceolate, margins pectinate-ciliate, usually glabrous on the back or a few hairs on the median sepal, tips subacute, often minutely denticulate; male sepals variably joined, sometimes almost free, subacute-denticulate, pilose at base of the lobes. Seeds uniformly brown at first, becoming longitudinally white-striate when wetted.

Two specimens diverge considerably from the main body of the species and have been described as a distinct variety.

Scapes glandular-pilose below the capitulum; sepals of the female flowers lanceolate, conspicuously pectinate-ciliate with hairs 0.25–0.45 mm long; seeds ellipsoid, 0.4–0.5 mm long · var. *wahlbergii*
Scapes glabrous below the capitulum; sepals of the female flowers narrowly lanceolate-oblong, the margins glabrous or inconspicuously ciliate with hairs 0.15 mm long; seeds narrowly cylindrical, 0.7–0.8 mm long · · · · · · · · · · · · · · var. *sinkabolensis*

Var. **wahlbergii**.

Zambia. B: Senanga Dist., 16 km N of Senanga, 31.vii.1952, *Codd* 7302 (K, PRE, SRGH). N: Mbala Dist., Lake Chilwa, 4.viii.1949, *Greenway* 8373 (K). C: Serenje Dist., Kundalila Falls, 24.viii.1983, *Parris & Croxall* 83/7 (K). S: Kalomo Dist., Machili, 2.vii.1963, *Fanshawe* 7898 (SRGH). **Zimbabwe**. W: Lupane Dist., roadside, 17.viii.1963, *Bingham* 833 (K, SRGH). C: Harare Dist., no locality, 3.x.1945, *Wild* 154 (K, SRGH). E: Nyanga Dist., Mare R., 21.x.1946, *Wild* 1551 (K, SRGH). **Malawi**. S: Zomba Plateau, between Chiradzulu and Malumbe, 10.x.1971, *Jordan* 5005 (K).

Also in South Africa, Angola and northwards to Uganda, Ethiopia, Central African Republic and Nigeria; the most widespread African species in the genus. Marshy places in open grassland, open parts of peat bogs, and the swampy or muddy margins of lakes, rivers and streams, the leaf rosettes sometimes submerged in shallow water, locally dominant; 1000–1850 m.

Conservation notes: Widely distributed; not threatened.

A small perennial species forming clusters of woolly-centred rosettes in bogs and marshes. The golden-brown capitula on glandular scapes are distinctive. The leaves have markedly incurving tips in most specimens from Zambia southwards, but in more northern populations (*S. chevalieri*) the leaves simply curve outwards without upturning tips.

A few specimens have paler yellow capitula, and can be difficult to distinguish from *S. longibracteatus*, especially if the perennial base is not well developed. *S. wahlbergii* has a shorter, more spathe-like limb to the sheath than *S. longibracteatus*, and usually a succession of capitula of different ages on scapes of very different heights within one rosette.

Var. **sinkabolensis** S.M. Phillips in Kew Bull. **52**: 73–89 (1997); in Kirkia **17**: 59 (1998). Type: Zambia, Mwinilunga Dist., Sinkabolo dambo, 20.x.1937, *Milne-Redhead* 2861 (K holotype).

Zambia. W: Mwinilunga Dist., Sinkabolo dambo, 20.x.1937, *Milne-Redhead* 2861 (K). **Zimbabwe**. C: Marondera Dist., Digglefold, 24.x.1949, *Corby* 515 (K, SRGH).

2. **Syngonanthus poggeanus** Ruhland in Engler, Pflanzenr. **13**: 247 (1903). —Hess in Ber. Schweiz. Bot. Ges. **65**: 190 (1955). —Kimpouni in Fragm. Flor. Geobot. **37**: 141 (1992). —Phillips in Kirkia **17**: 59 (1998). Type: Angola, Lunda, Mona Quimbundo, viii.1876, *Pogge* 457 (B holotype).

Perennial forming clusters of leaf-rosettes from a tough rootstock. Leaves linear, 1.5–8 cm long, 1–2.5 mm wide, thick, the young leaves usually densely pilose with pointed and glandular hairs, the older leaves glabrescent, outwardly curving, tapering to an acute tip. Scapes 1–10 or more, up to 40(60) cm high, 3-ribbed, villous with pointed hairs intermixed with short glandular hairs, densely so below the capitulum, appressed between the ribs lower down, glabrescent; sheaths equalling or exceeding the leaves, densely to sparsely pilose with glandular hairs, shortly slit with an appressed, abruptly acute limb. Capitulum whitish, (5)7–8 mm wide, hemispherical to globose, appearing glabrous; involucral bracts mostly shorter than the capitulum width, thinly scarious and pallid from a pale yellowish coriaceous base, obtuse-lacerate, the outermost oblong, 1.5–1.8 mm long, the inner progressively longer, up to 2.5–3 mm long, narrowly elliptic-oblong with a few short marginal cilia, innermost bracts translucent, linear-lanceolate, acuminate, the lower margins ciliate; receptacle villous. Flowers 2–2.2 mm long, pallid; sepals of female flowers narrowly lanceolate-oblong, acute or obtuse-denticulate, shortly ciliate on lower margins and sometimes sparsely along the midline; stigmas with alternating swollen-tipped appendages; male sepals resembling the female. Seeds narrowly oblong, 0.7 mm long, brown with indistinct longitudinal stripes that eventually turn white.

Zambia. B: Kalabo Dist., c.3 km W of Kalabo, 13.xi.1959, *Drummond & Cookson* 6443 (K, SRGH). N: Kawambwa Dist., Ntumbachushi Falls, 19.iv.1989, *Radcliffe-Smith, Pope & Goyder* 5736 (K). W: Mwinilunga Dist., c.45 km E of Mwinilunga, 11.ix.1930, *Milne-Redhead* 1088 (K).

Also in Angola, Congo and Burundi. Swampy grassland on sandy soils, and wet ground bordering rivers and waterfalls; 1100–1800 m.

Conservation notes: Widely distributed; not threatened.

This is the largest *Syngonanthus* species in Africa, and is often confused with *S. angolensis* (for differences see under that species).

Greenway 5403 (Ishiba Ngandu, 19.vii.1938 (K, SRGH)) is similar to *S. poggeanus* but has yellowish capitula and somewhat inflated sheath-tips, perhaps due to introgression from *S. wahlbergii*. It differs from both these species by the pilose backs to the sepals in the female flowers.

3. **Syngonanthus angolensis** H. Hess in Ber. Schweiz. Bot. Ges. **65**: 193 (1955). — Phillips in F.T.E.A., Eriocaulaceae: 36 (1997). —Phillips in Kirkia **17**: 601 (1998). Type: Angola, Cuando Cubango, R. Luassinga, 60 km E of Menongue, 1400 m, 28.i.1952, *Hess* 52/2098 (Z holotype). FIGURE 13.4.**9**.

Tufted perennial from a short rootstock. Leaves narrowly linear, loosely erect or curving outwards, 1.5–7 cm long, 0.6–1.7 mm wide, pilose on both surfaces especially when young, glabrescent, acute. Scapes up to 8, 20–45 cm high, 3-ribbed, pilose with pointed hairs, densely so below the capitulum, appressed-pilose between the ribs lower down, glandular hairs absent or sparse; sheaths equalling or longer than the leaves, softly pilose to subglabrous, the limb scarcely inflated, rather stiffly erect, tapering to an acute tip. Capitulum 7–9.5 mm wide, ivory-white to cream-coloured with a brown base and brown inside towards the centre; involucral bracts mostly shorter than the capitulum width, the longest reaching the periphery, brown near the scape, pallid around periphery, subacute to acute, the outermost narrowly ovate, 2–2.3 mm long, glabrous, the inner narrowly elliptic-oblong, 3–3.7 mm long, innermost shortly ciliate on the margins; receptacle villous. Flowers 2–2.5 mm long; sepals of female flowers pallid, narrowly oblong, ciliate on margins and back with fine non-persistent hairs, obtuse or acute; male sepals united and dark brown below the middle, the free lobes pallid, obtuse-denticulate. Seeds ellipsoid, 0.5 mm long, white-striate.

Zambia. N: Kawambwa Dist., Kale, Kalungwishi (Lwingishi) R., 16.iv.1963, *Symoens* 10274 (K); Chinsali Dist., Ishiba Ngandu (Shiwa Ngandu), 2.vi.1956, *Robinson* 1555 (K, SRGH). C: Serenje Dist., Bolelo R., 25.xii.1963, *Symoens* 10749 (K). **Malawi**. N: Mzimba Dist., c.5 km W of Mzuzu, 23.vii.1973, *Pawek* 7240 (K, SRGH).

Also in Angola, S Congo and S Tanzania. Boggy hollows and marshy areas adjoining watercourses; 1200–1800 m.

Conservation notes: Moderately widely distributed; not threatened.

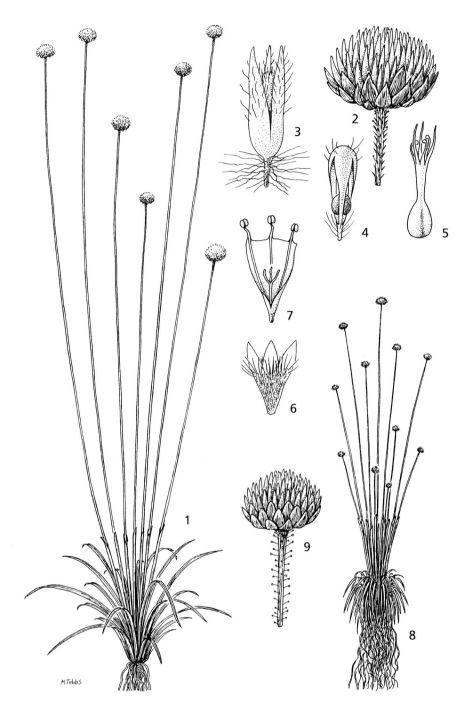

Fig. 13.4.**9**. SYNGONANTHUS ANGOLENSIS. 1, habit (× 2/3); 2, capitulum (× 4); 3, female flower (× 14); 4, female petals, joined and shrunk inwards at top, and ovary (× 14); 5, gynoecium (× 20); 6, male flower (× 14); 7, inside view of male petal-tube with stamens and vestigial ovary (not to scale). S. WAHLBERGII. 8, habit (× 2/3); 9, capitulum (× 4); 1–5 from *Milne-Redhead & Taylor* 10847, 6 & 7 from *Polhill & Paulo* 1524, 8 & 9 from *Drummond & Hemsley* 4637. Drawn by Margaret Tebbs. From Flora of Tropical Africa.

Most specimens of *S. angolensis* have pilose scapes with pointed hairs only, the hairs being particularly dense below the capitulum. However, short glandular hairs are occasionally also present below the capitulum intermixed with the longer pointed hairs, although they are never so obvious as in *S. poggeanus* or *S. longibracteatus*.

Corby 145 from Zimbabwe (C: Marondera Dist., Digglefold) has the pale capitula and pilose female sepals of this species, but the small recurving leaves, thinly glandular-pilose scapes and spathe-like limb to the sheath are closer to *S. wahlbergii*. Its status cannot be determined with certainty at present.

4. **Syngonanthus robinsonii** Moldenke in Phytologia **17**: 437 (1968). —Phillips in Kirkia **17**: 60 (1998). Type: Zambia, Mporokoso Dist., 55 km ESE of Mporokoso, Kasanshi Dambo, 13.v.1962, *Robinson* 5167 (NY holotype, K, M, SRGH).

Slender rosulate annual. Leaves numerous, subulate, recurving, 0.5–2.5 cm long, 0.3–1 mm wide, appressed-hispid, obtuse. Scapes 1–10, 12–25 cm high, 3-ribbed, glandular-pilose with spreading hairs; sheaths longer than the leaves, tight, glandular-pilose, the limb acuminate. Capitulum shiny white, 4–6(8) mm wide, densely hairy; involucral bracts creamy-buff to light golden, concave, outermost small, half as wide as the capitulum or less, 1.1–1.3 × 0.5–0.6 mm, lanceolate to ovate, ± glabrous with only a few marginal hairs, darker coloured than the inner, obtuse, successive bracts longer and paler with more hairs, the innermost narrowly elliptic and extending to the capitulum periphery, 2–2.8 × 0.6–0.7 mm, white, subacute, densely pilose on margins and back above the middle, capitulum base thus coloured yellowish near the scape, and white and fluffy around the periphery. Flowers 1.4–2 mm long; sepals of both male and female flowers villous in the middle third with spreading white hairs on the back and also inside; female sepals acute, male sepals obtuse-denticulate. Seeds plumply ellipsoid, 0.35 mm long, uniformly brown.

Zambia. N: Luwingu Dist., Chishinga Ranch, near Luwingu, 27.iv.1961, *Astle* 543 (K, SRGH); Kawambwa Dist., Ntumbachushi (Timnatushi) Falls, 19.iv.1957, *Richards* 9340 (K).

Only found in N Zambia and neighbouring parts of Congo. Damp sand, often among rocks or near waterfalls, frequently forming large colonies; 1250–1600 m.

Conservation notes: Localised distribution; possibly Vulnerable.

The species can be recognised by its slender annual habit and hairy capitula, the short, soft involucral bracts providing a brown base with a white marginal band to the capitulum. The seeds, which lack white stripes, are very unusual in African *Syngonanthus*, and will confirm the identification.

5. **Syngonanthus mwinilungensis** S.M. Phillips in Kew Bull. **52**: 81 (1997); in Kirkia **17**: 61 (1998). Type: Zambia, Mwinilunga, 7 km N of Kalene Hill, 16.iv.1965, *Robinson* 6586 (K holotype, SRGH). FIGURE 13.4.**10**.

Slender annual. Leaves many in a small basal rosette, linear, recurving, c.1 cm long, 0.3–0.6 mm wide, appressed-hispid, narrowly obtuse. Scapes 2–7, 17–20 cm high, 3-ribbed, thinly glandular-hairy with spreading hairs; sheaths longer than the leaves, 1.7–2.5 mm long, tight, glandular-pilose. Capitulum ivory-white throughout (at most a faint yellowish coloration around the scape), 5 mm wide, woolly within from receptacular hairs but only thinly hairy at the periphery; involucral bracts in several series of increasing length, tips of the longest slightly exceeding the flowers, stiffly scarious, sharply acute, the outermost 1.5 × 0.8 mm, increasing to 2.7 × 0.7 mm, margins ciliate especially the inner, only the innermost with a few hairs on the upper back. Flowers 2 mm long; sepals of both male and female flowers acuminate, ciliate on the margins, pilose on the back and also inside. Seeds ellipsoid, 0.4 mm long, uniformly blackish-brown, the ovary appearing blackish.

Zambia. W: Mwinilunga Dist., Zambezi Rapids, 18.v.1969, *Mutimushi* 3296a (SRGH).

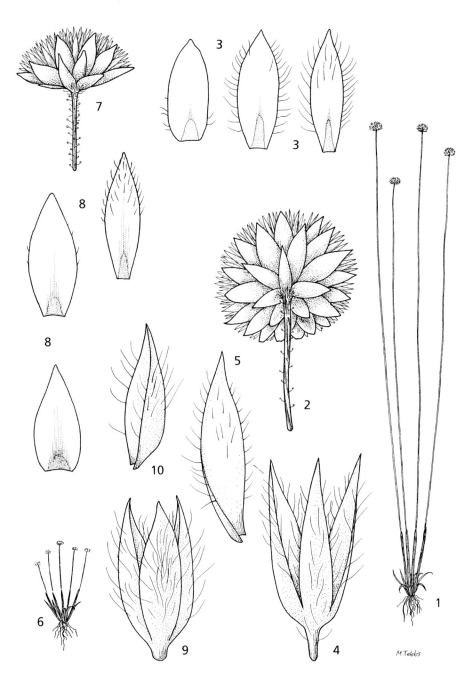

Fig. 13.4.**10**. SYNGONANTHUS MWINILUNGENSIS. 1, habit (× 1); 2, capitulum (× 15); 3, involucral bracts (× 24); 4, female flower (× 45); 5, female sepal (× 45). S. EXILIS. 6, habit (× 1); 7, capitulum (× 15); 8, involucral bracts (× 24); 9, female flower (× 45), 10, female sepal (× 45), 1–5 from *Robinson* 6586, 6–10 from *Milne-Redhead* 2653. Drawn by Margaret Tebbs. From Kew Bulletin.

Not known elsewhere. Damp sandy soil over rock; c.1400 m.

Conservation notes: Apparently endemic to this part of NW Zambia. Very localised; probably Vulnerable.

This species is closely related to *S. robinsonii* but is distinguished mainly by its much stiffer, sharply acute, white involucral bracts radiating across the capitulum base. The flowers have the same structure as those of *S. robinsonii*, with acute sepals pilose on the inside as well as the back, but are slightly larger. The uniformly brown, non-striate seed is enclosed by a blackish ovary wall, easily visible through the translucent, pallid sepals of the female flower.

6. **Syngonanthus longibracteatus** Kimpouni in Bull. Jard. Bot. Belg. **61**: 339 (1991). —Phillips in F.T.E.A., Eriocaulaceae: 38 (1997); in Kirkia **17**: 61 (1998). Type: Congo, Katanga, Kundelungu Plateau, 8 km NW of Lualala near source of R. Nungwe, *Lisowski* 58176 (POZG holotype, BR).

Slender annual. Leaves many in a dense basal rosette, linear to subulate, 0.5–2 cm long, 0.5–1 mm wide, pilose with glandular and pointed hairs, subacute. Scapes up to c.20 but often much fewer, 8–25 cm high, 3-ribbed, pilose with spreading glandular hairs intermixed with appressed pointed hairs; sheaths equalling or often longer than the leaves, conspicuously glandular-hairy. Capitulum shiny creamy-white or slightly yellowish, 5–6 mm wide; involucral bracts pale yellowish to light golden-brown, the outermost ovate-oblong and subacute, 1.4–1.7 × 0.6–0.8 mm, grading to oblong, the longest equalling the flowers, 2.3–2.8 × 0.7–0.9 mm, mostly glabrous, the innermost ciliate on the margins. Flowers 1.4–1.7 mm long; sepals of female flowers lanceolate, acute, pectinate-ciliate along the margins with hairs c.0.4 mm long, the centre back usually thinly hairy, occasionally glabrous; sepals of male flowers similar, acute, basal united portion light gold, the free lobes pallid. Seed brown with longitudinal white stripes.

Zambia. N: Mbala Dist., Dhulmiti Kloof, 12.v.1955, *Richards* 5678a (K). C: Kabwe Dist., 24 km N of Kabwe (Broken Hill), on W side of Great North Road, just beyond Mulungushi R., 23.ix.1947, *Brenan & Greenway* 7923 (K). **Zimbabwe**. N: Guruve Dist., Nyamunyecho Estate, Karoi vlei, 8.x.1978, *Nyariri* 391 (K, SRGH). C: Harare (Salisbury), 7.v.1931, *Brain* 3770 (K, SRGH). S: Bikita Dist., W bank of Turgwe R., Turgwe-Dafana confluence, 5.v.1969, *Biegel* 3020 (K, SRGH). **Mozambique**. N: Lago Dist., Maniamba, by rio Messinge, 900 m, 13.ix.1934, *Torre* 225 (LISC).

Also in S Congo and Tanzania. Ditches, streamsides and wet depressions on sandy soil; 900–1500 m.

Conservation notes: Widely distributed; not threatened.

Distinguished from similar annual species by its hairy scapes with a mixture of glandular and pointed hairs, and by the long, oblong involucral bracts. Specimens from Zimbabwe have often been named as *S. wahlbergii*. Usually the habit difference is sufficient to distinguish them, *S. wahlbergii* being a perennial forming clusters of rosettes from a branching rhizome. However, where the habit is not clear there can be difficulty as both have similar-sized coloured capitula on glandular-pilose scapes. The capitula of *S. longibracteatus* are paler than the golden-brown ones of *S. wahlbergii* and are usually fewer of more equal height. The leaves are also broader in *S. longibracteatus* which lacks the flared, spathe-like limb to the sheaths so characteristic of *S. wahlbergii*.

7. **Syngonanthus exilis** S.M. Phillips in Kew Bull. **52**: 85 (1997); in Kirkia **17**: 63 (1998). Type: Zambia, Mwinilunga, Kalenda dambo, 8.viii.1938, *Milne-Redhead* 2653 (K holotype). FIGURE 13.4.**10**.

Dwarf rosulate herb from a short thickened stem, probably a short-lived perennial. Leaves linear, 0.5–1 cm long, 0.3–0.7 mm wide, appressed-hispid with thick white hairs on the upper surface, acute. Scapes 1–4, 2.5–4 cm high, 3-ribbed, pilose with pointed and glandular hairs intermixed; sheaths glandular-pilose, the limb subacute, appressed to the scape, about $^1/_4$ of total sheath length. Capitulum whitish, 4–4.5 mm wide, obviously pilose; involucral bracts in 2–3 series, pale buff, lanceolate, acute, subequal, the outermost only slightly shorter than the inner, tips of the inner slightly protruding beyond the flowers, outermost 1.8 × 0.8 mm, glabrous, the inner 2.2 × 0.7 mm, ciliate on margins and upper back; receptacle villous with long hairs almost as long as the flowers. Flowers 0.4–0.8 mm long; sepals of female flowers lanceolate, acute, pilose on the margins and back with hairs 0.5–0.6 mm long; stylar appendages present; male sepals joined only at the base, ciliate on the margins, densely pilose on the back in the middle third. Seeds longitudinally white-striate.

Zambia. W: Mwinilunga Dist., Kalenda dambo, 8.viii.1938, *Milne-Redhead* 2653 (K). Known only from the type collection. Wet peaty mud near exposed laterite; c.1300 m.
Conservation notes: Only known from the type, with no recent collections; Data Deficient, possibly Endangered.

This small species has capitula and flowers very similar to those of *S. longibracteatus*, but the involucral bracts do not contrast quite so much with the whitish flowers and are slightly more lanceolate than lanceolate-oblong. The scapes are much shorter, and arise from an untidy tuft of loosely erect leaves from a basal rootstock, in contrast to the neat annual rosette of recurving leaves of *S. longibracteatus*.

8. **Syngonanthus paleaceus** S.M. Phillips in Kew Bull. **53**: 491 (1998); in Kirkia **17**: 64 (1998). Type: Zambia, 45 km N of Mansa, 5.vi.1960, *Robinson* 3735 (SRGH holotype).

Dwarf rosulate herb, probably annual. Leaves many, linear-subulate, c.1 cm long, 0.5–0.7 mm wide, recurving. Scapes up to 50 or more, 1–6 cm high developing in close succession, slender, 3-ribbed, pilose with spreading glandular hairs intermixed with some shorter pointed ones; sheaths densely glandular-pilose. Capitulum 3–3.8 mm wide; involucral bracts straw-coloured, in 2–3 series, slightly longer than the flowers, 1.4–1.7 mm long, stiffly scarious, acute, the outer ovate, glabrous, the inner narrower, ciliate on the margins; receptacle pilose. Flowers c.1 mm long, outermost female, male flowers few; sepals golden-brown, petals white. Sepals of female flowers narrowly lanceolate-oblong, pectinate-ciliate on the margins, acute; male sepals obtuse-denticulate, thinly pilose on the centre back; style with 3 filiform stigmas but only one well-developed swollen-tipped appendage. Mature seed not seen.

Zambia. N: Mansa Dist., 45 km N of Mansa (Fort Rosebery), 5.vi.1960, *Robinson* 3735 (SRGH).
Known only from the type, collected in a dambo; 1250 m.
Conservation notes: Only known from the type; Data Deficient, probably Endangered.

9. **Syngonanthus welwitschii** (Rendle) Ruhland in Engler, Pflanzenr. **13**: 248 (1903). —Hess in Ber. Schweiz. Bot. Ges. **65**: 197 (1955). —Kimpouni in Fragm. Fl. Geobot. **37**: 152 (1992). —Phillips in Kirkia **17**: 64 (1998). Type: Angola, Huíla, near Lopollo, iv–v.1860, *Welwitsch* 2447 (BM holotype, B, K, Z).
    Paepalanthus welwitschii Rendle in Cat. Afr. Pl. Welw. **2**: 102 (1899). —N.E. Brown in F.T.A. **8**: 262 (1901).

Tiny rosulate annual. Leaves numerous, narrowly linear-subulate, less than 1 cm long, c.0.5 mm wide, glabrous or thinly pilose, subacute. Scapes up to 50, 1–4 cm high, filiform, 3-ribbed, pilose with long, spreading, glandular hairs intermixed with shorter pointed ones; sheaths glandular-pilose, limb $^2/_3$–$^3/_4$ as long as the closed tubular lower portion, its tip often recurving. Capitulum 1.5–2.5 mm wide, few flowered; involucral bracts few, subequal, shiny white, extending beyond the flowers, 1.2–1.5 mm long, lanceolate or ovate, acute, the outer glabrous or pilose, the inner pectinate-ciliate on the margins; receptacle densely pilose with long straight hairs. Flowers 5–9, 0.5–0.7 mm long, 4–7 subsessile female flowers surrounding 1–2 shortly pedicelled male flowers; female sepals straw-coloured to brown, oblong, pectinate-ciliate on the margins, obtuse-denticulate; style with 3 long stigmatic branches but no alternating swollen appendages. Male sepals similar to the female, thinly pilose. Seeds 0.35 mm long, reddish-brown, faintly longitudinally striate, the stripes becoming white after wetting.

Zambia. B: Zambezi Dist., 21 km W of Zambezi (Balovale) pontoon, 26.v.1960, *Angus* 2277 (SRGH). **N**: Mbala Dist., Kambole, Katenga Falls, 5.vi.1957, *Richards* 10009 (K); Mbala Dist., 20 km from Chikwanda, 25.vi.1963, *Symoens* 10468b (K).

Also in Tanzania, Angola, S Congo and Sierra Leone. Wet ground in grassy and weedy places, on sandy or gritty soils; 1500–1750 m.

Conservation notes: Moderately widely distributed; not threatened.

Easily distinguished from other species by its very small size, with tiny few-flowered capitula. The exceptionally long limb to the scape-sheath is a good confirmatory character. The absence of stylar appendages alternating with the 3 stigmas is another unusual feature of this species, but difficult to see without a microscope.

The only other African species lacking stylar appendages is *S. schlechteri* Ruhl., known from Congo, Gabon and Tanzania. This has larger white capitula 3.5–4.5 mm wide containing more flowers, and set on scapes up to 15 cm high.

4. PAEPALANTHUS Mart.

Paepalanthus Mart., Ann. Sci. Nat., Bot., sér.2, **2**: 28 (1834) nom. cons.[2] —Ruhland in Engler, Pflanzenr. **13**: 121–223 (1903).

Variable herbs, annual or perennial, the stem short or elongate; leaves thin to thick and coriaceous; capitula single or arranged in umbels. Capitulum villous; floral bracts present. Flowers 2–3-merous; petals eglandular. Male flowers: sepals free except at the extreme base; petals united into a glabrous funnel-shaped tube bearing the stamens on its truncate upper margin, stamens as many as petals, white, petal-tube finally rolled inwards and enclosing the stamens, a rudimentary pistil present. Female flowers: sepals free except at extreme base and usually becoming rigid at maturity; petals free; style with glandular appendages alternating with the stigmas, these simple or bifid. Seeds variable, often with longitudinal ridges.

About 500 species, almost entirely confined to tropical America.

Paepalanthus is the largest genus in the family and is very heterogeneous, both vegetatively and in floral characters. Only two species occur in Africa, the widespread *P. lamarckii* and *P. pulvinatus* N.E. Br., endemic to Sierra Leone.

Paepalanthus lamarckii Kunth, Enum. Pl. **3**: 506 (1841). —Ruhland in Engler, Pflanzenr. **13**: 160 (1903). —Milne-Redhead in Kew Bull. **3**: 472 (1948). — Phillips in F.T.E.A., Eriocaulaceae: 39 (1997); in Kirkia **17**: 65 (1998). Type: Guyana (Guinea), *Willd. Herb.* 2372 (B holotype). FIGURE 13.4.11.

Small annual, developing with age an unbranched wiry stem clothed in old leaves and fibrous roots, the new leaves forming a loose rosette at the stem tip. Leaves linear to lanceolate, 2–3 cm long, 1–3 mm wide, spongy, scattered-pilose to glabrescent, the tip hardened, subacute. Scapes many in a terminal tuft, 2–7 cm high, 3-ribbed; sheaths shorter than the leaves, loose with an acute limb. Capitulum 3–4 mm wide, subglobose with intruded base at maturity, grey, densely

---

[2] Taxon **47**: 743–744 (1998) and Taxon **49**: 274 (2000).

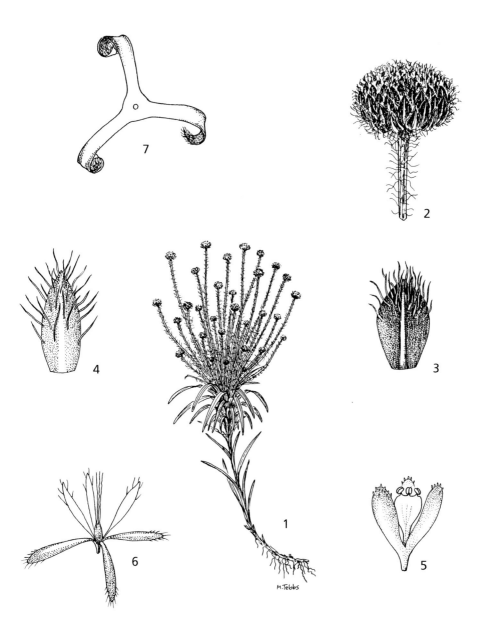

Fig. 13.4.**11**. PAEPALANTHUS LAMARCKII. 1, habit (× ²/₃); 2, capitulum (× 4); 3, involucral bract (× 20); 4, floral bract (× 20); 5, male flower (× 20); 6, female flower (× 20); 7, recurved female calyx after shedding petals and ovary (× 40), all from *Goyder, Pope & Radcliffe-Smith* 3081. Drawn by Margaret Tebbs. From Flora of Tropical East Africa.

villous, the immature central flowers completely obscured by white hairs; involucral bracts lanceolate to lanceolate-oblong, 1–1.2 mm long, firmly membranous, brownish-grey with a paler central stripe, villous with coarse spreading hairs from the margins and back, acute; floral bracts angular-obovate, dark grey with a paler central stripe, coarsely villous above the middle; flowers trimerous, 0.8–1 mm long. Male flowers: sepals oblong-spathulate, concave, dark grey with paler central stripe, densely pilose at the subacute tips; petal-tube borne on a stipe; vestigial gynoecium represented by 3 elongate glands. Female flowers: sepals resembling the male at first; petals translucent, equalling sepals, narrowly oblong-spathulate, scattered-pilose; at maturity the sepals hardening, recurving and raising the petals and ovary to the capitulum surface. Seeds ellipsoid, 0.3 mm long, light brown.

Zambia. N: Kawambwa Dist., Muchinga Escarpment, Mbereshi Nat. Forest, 15 km W of Kawambwa, 19.iv.1989, *Goyder, Pope & Radcliffe-Smith* 3081 (K).

Scattered localities throughout W Africa, and in Congo, Tanzania (Mafia Is.) and Madagascar. Widespread in tropical America from Belize to Brazil and the West Indies. Around the drying margins of temporary pools and open places on damp sand; c.1100 m.

Conservation notes: Widespread distribution; not threatened.

The method of seed dispersal is distinctive and unlike any method found in African Eriocaulaceae. The grey and white striped, hardened female sepals coil outwards, hence raising the petals and ovary to the capitulum surface where they are shed (Lecomte in J. Bot., sér.2 **1**: 136, 1908). The 3-armed, empty calyces with a central hole remain clinging to the capitulum and are a good character for distinguishing this species from *Eriocaulon* without the need for dissection. The dry, brown, fibrous roots and villous scapes also serve to distinguish it from *Eriocaulon* without examining details of the tiny flowers.

TYPHACEAE

by L. Catarino & E.S. Martins

Monoecious perennial, aquatic or marsh herbs with creeping rhizomes. Stems erect, unbranched, terete, terminated by dense cylindrical flower-spikes. Leaves mostly radical, in two rows, sheathing at base, elongate-linear, parallel-veined. Inflorescence a dense double spike of closely packed flowers with male flowers in the upper portion, contiguous to or separated from female flowers in the lower portion, in a long peduncle. Flowers minute and numerous. Male flowers ephemeral, usually subtended by variously shaped scales or bracteoles; perianth absent or of 3–6 small scales; stamens 1–5; filaments free or variously joined; anthers linear, basifixed, opening by longitudinal slits. Female flowers without bracteoles or with slender clavate or spathulate bracteoles, abortive clavate female flowers (carpodia) frequently produced; perianth of several fine, persistent, filiform or clavate hairs. Ovary superior, often stalked, fusiform, unilocular with a solitary pendulous ovule; style elongated, slender; stigma linear or lanceolate. Fruits minute, ellipsoid or subcylindrical. Seed with a striate testa and mealy endosperm; embryo narrow, nearly as long as the seed.

A family with a single cosmopolitan genus and about 15 species of shallow freshwater, marshes and wet soils, with 4 species in tropical Africa. Hybrids frequently occur. Often gregarious and dominant over large areas, stands of these rushes often act as a refuge for wildlife. Juveniles are normally submerged and adult plants are emergent.

In southern Africa the ash is sometimes used as a source of salt, the rhizomes and immature inflorescences are sometimes eaten, and the leaves have been used in weaving. Sometimes populations became weeds in irrigation systems and rice fields.

TYPHA L.

Typha L., Sp. Pl.: 971 (1753); Gen. Pl. ed.5: 418 (1754).

Description as for the family.

Female flowers with bracteoles; stigma linear; male and female spikes (1)2–7 cm distant · **1.** *domingensis*
Female flowers without bracteoles; stigma lanceolate or spathulate; male and female spikes contiguous or up to 1(2) cm distant · · · · · · · · · · · · · · · · · · **2.** *capensis*

1. **Typha domingensis** Pers., Syn. Pl. **2**: 532 (1807). —Napper in F.T.E.A., Typhaceae: 2, fig.1 (1971). —Cook, Aq. Wetl. Pl. Sthn. Africa: 258 (2004). Type: Dominican Republic (St Domingue), unknown collector (?L holotype). FIGURE 13.4.**12**.

    Typha australis Schumach. & Thonn. in Schumacher, Beskr. Guin. Pl.: 401 (1827). —N.E. Brown in F.T.A. **8**: 135 (1901). —Mogg in Macnae & Kalk, Nat. Hist. Inhaca Is., Moçamb.: 139 (1958). —Adam in F.W.T.A., ed.2, **3**: 131 (1968). —Binns, First Check List Herb. Fl. Malawi: 101 (1968). Type: Ghana, *Thonning* s.n. (C holotype).

    Typha latifolia subsp. *capensis* sensu Munday & Forbes in J. S. Afr. Bot. **45**: 2 (1979).

Perennial herb, robust, up to 3 m tall, glabrous. Stems erect, simple, terete, connected by rhizomes. Leaves distichous, not differentiated into petiole and blade, the underground ones scale-like, the aerial linear, sheathing at the base, tapering to an obtuse apex, convex on the abaxial surface and flat or slightly concave on the adaxial surface, thinner and flat towards the apex, up to 150 cm or more long, (5)7–10(15) mm broad and 2–4(5) mm thick, margins smooth, often undulated in the upper half, green. Inflorescence a dense cylindrical brown or yellowish-brown double spike; upper male portion (15)20–30(35) cm long and 8–12(13) mm in diameter; lower female portion (10)15–25(28) cm long and (5)8–20(25) mm in diameter, (1)2–6(7) cm distant. Male flowers ephemeral, usually with variously shaped, linear, lanceolate, cuneate or forked, brownish bracteoles; stamens with slender filaments bearing 1 to 4 anthers; anthers 4–5 mm long, linear, basifixed, often distinctly twisted, 2-celled, opening longitudinally, with the apex swollen, dark brown; pollen simple or rarely compound. Female flowers with brownish, lanceolate or clavate bracteoles, perianth of several, slender, colourless, simple hairs, ovary superior, stalked, often brown-mottled, narrow, unilocular with a solitary pendulous ovule; style linear, brown; abortive female flowers (carpodia) frequent, with brown, clavate stigma. Fruits minute, stalked, elipsoid or subcylindrical, c.1 × 0.5 mm wide at the middle.

    Botswana. N: Okavango swamps, Zibadianga lagoon, 1000 m, fl.& fr. 20.xi.1972, *Gibbs Russell* 2156 (K). **Zambia**. B: Senanga Dist., Kaunga, near Kwando (Mashi) R., fl. 18.x.1962, *Reynolds* 86 (K, SRGH). N: Kaputa Dist., shoreline of Lake Mweru Wantipa, N of Kampinda village, 870 m, fl.& fr. 25.vii.1962, *Tyrer* 114 (BM). E: Katete, 80 km W of Chipata (Fort Jameson), 1050 m, fl.& fr. 14.vi.1954, *Robinson* 877 (K). S: Monze, Lochinvar Ranch, fl. 8.viii.1963, *van Rensburg* 2402 (K). **Malawi**. N: Karonga Dist., 45 km N of Chilumba, c.500 m, fl.& fr. 4.vii.1970, *Pawek* 3576 (K). S: Zomba Dist., Lake Chilwa (Shirwa), 520 m, fl.& fr. 22.x.1941, *Greenway* 6345 (K). **Mozambique**. N: Mueda Dist., 19 km from Chomba on road to Negomano, c.600 m, fl.& fr. 13.iv.1964, *Torre & Paiva* 11886 (K, LISC, LMU, SRGH). Z: Morrumbala Dist., Tembe-Tembe (Vila Bocage), Águas Quentes, fl. 3.x.1944, *Mendonça* 2333 (LISC, LMA). T: Cahora Bassa Dist., Songo, near Zambezi R., fr. 5.xi.1973, *Macedo* 5356 (LISC, LMA). MS: Beira Dist., N of Macuti beach, fl.& fr. 10.ix.62, *Noel* 2486 (K, LISC). GI: Massinga Dist., Rio das Pedras, fl.& fr. 27.x.1947, *Barbosa* 578 (COI). M: Maputo, between Costa do Sol and Marracuene, fl.& fr. 10.xii.1979, *de Koning* 7733 (BM, K, LMU).

    Pantropical, including South Africa. Often dominant in fresh and brackish swamps, backwaters, lagoons, small dams and along watercourses; up to 1100 m (2250 m outside the Flora area).

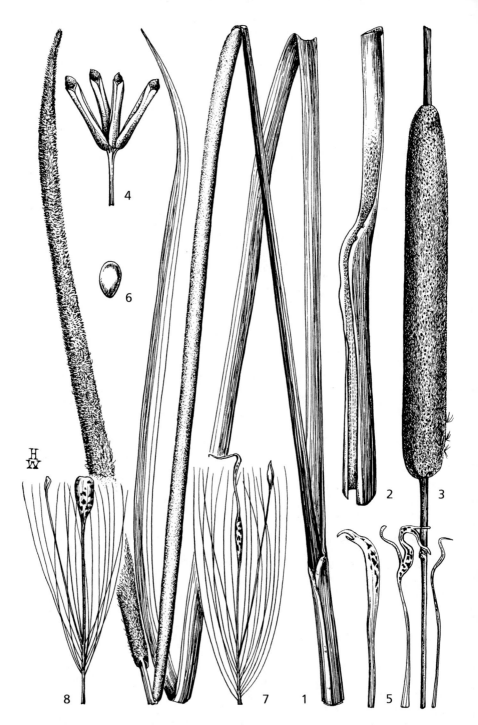

Fig. 13.4.**12**. TYPHA DOMINGENSIS. 1, part of shoot showing mature male and immature female inflorescences (× ¹/₂); 2, leaf sheath (× ¹/₂); 3, mature female spike (× ¹/₂); 4, male floret (× 7¹/₂); 5, male bracteoles (× 13); 6, pollen (× 100); 7, female floret with ovary and bracteole (× 7¹/₂); 8, sterile female floret with carpodium and bracteole (× 7¹/₂). 1–2, 4–8 from *Greenway & Kanuri* 12567, 3 from *Tanner* 1135. Drawn by Heather Wood. From Flora of Tropical East Africa.

Conservation notes: Widespread and common species.

Galen Smith (Archiv. für Hydrobiologie, Beiheft **27**: 129–139, 1987) suggests that *T. domingensis* should be treated as a single highly-variable worldwide species of tropical and warm-temperate regions. According to Cook (2004), this species is normally larger and more robust than *T. capensis*.

2. **Typha capensis** (Rohrb.) N.E. Br. in Fl. Cap. **7**: 32 (1897). —Rendle in Cat. Afr. Pl. Welw. **2**(1): 85 (1899). —N.E. Brown in F.T.A. **8**: 136 (1901). —Napper in F.T.E.A., Typhaceae: 4 (1971). —Clarke & Klaassen, Water Pl. Namibia: 62 (2001). —Cook, Aq. Wetl. Pl. Sthn. Africa: 259 (2004). Type: South Africa, Western Cape, Uitenhage, Zwartkops R., *Ecklon & Zeyher* 913 (SAM lectotype).

> *Typha latifolia* L. subsp. *capensis* Rohrb. in Verh. Bot. Vereins Prov. Brandenburg **11**: 96 (1870). —Anderson in F.S.A. **1**: 54, fig.13 (1966). —Roessler in Merxmüller, Prodr. Fl. SW Afrika, fam.164: 1 (1967). —Binns, First Check List Herb. Fl. Malawi: 101 (1968). —Ross, Fl. Natal: 54 (1972). —Gibbs Russell in Kirkia **10**: 433 (1977). —Munday & Forbes in J. S. Afr. Bot. **45**: 2 (1979).
>
> *Typha latifolia* sensu Engler, Pflanzenw. Ost-Afr. **C**: 93 (1895). —N.E. Brown in F.T.A. **8**: 136 (1901). —Napper in F.T.E.A., Typhaceae: 5, fig.1/9 (1971).
>
> *Typha australis* sensu N.E. Brown in Fl. Cap. **7**: 31 (1897). —Eyles in Trans. Roy. Soc. S. Afr. **5**: 292 (1916), non Schumach. & Thonn.

Perennial herb, robust, glabrous, up to 3 m tall. Stems erect, simple, terete, connected by rhizomes. Leaves in two rows, not differentiated into petiole and blade, the underground ones scale-like, the aerial ones linear, sheathing at the base, tapering to an obtuse apex, convex on the abaxial surface and flat or slightly concave on the adaxial surface, thinner and flat towards the apex, up to 200 cm long, 5–13 mm broad and 2–3(5) mm thick, margins smooth, often undulated in the upper half, green. Inflorescence a dense cylindrical brown or yellowish-brown double spike; upper male portion (8)10–20(28) cm long and (7)8–15(20) mm in diameter; lower female portion (7.5)10–20(25) cm long and (6)10–20(25) mm in diameter, contiguous with the male or up to 1(2) cm distant. Male flowers ephemeral, usually with a few variously shaped, linear, lanceolate, cuneate or forked, brownish bracteoles; stamens with slender filaments bearing 1–4 anthers; anthers 4–5 mm long, linear, basifixed, often distinctly twisted, 2-celled, opening longitudinally, with apex swollen and dark brown; pollen simple or rarely compound. Female flowers without bracteoles, perianth of several slender, colourless simple hairs; abortive female flowers (carpodia) often present. Ovary superior, stalked, often brown-mottled, narrow, unilocular with a solitary pendulous ovule; style lanceolate to spathulate, brown. Fruits minute, stalked, elipsoid or subcylindrical, c.1 × 0.5 mm wide.

Botswana. SE: Gaborone, Gaborone Dam, 950 m, fl.& fr. 18.i.1974, *Moot* 122C (K). **Zambia**. B: Kaoma Dist., Luena R., Mangango Mission, fr. 17.iv.1964, *Verboom* 1730 (K). C: Lusaka Dist., c.10 km E of Lusaka, 1300 m, fl.& fr. 7.xi.1972, *Strid* 2469 (K). **Zimbabwe**. N: Kariba Dist., S bank of Lake Kariba, near islands 23 & 24, c.460 m, fl.& fr. xi.1964, *Mitchell* 1074 (K, LISC). W: Bulawayo Dist., edge of Hillside Dam, 1370 m, fl.& fr. i.1958, *Miller* 4961 (K, LISC). C: Harare Dist., Ruwa R., fl. xi.1951, *Drummond* 4940 (K, LISC). E: Nyanga Dist., Tsonzo Division, Nyarandi R., fl. 1.xi.1950, *Chase* 3068 (BM, K, LISC). S: Chiredzi Dist., Hippo Valley, fl.& fr. 27.iii.1971, *Taylor* 163 (K). **Malawi**. N: Mzimba, Mbowe Dam, 16 km SW of Mzuzu, fl.& fr. 22.xii.1968, *Pawek* 1610 (K). **Mozambique**. N: Nacala-Velha Dist., between Mingúri (Fernão Veloso) and Itoculo, fl.& fr. 15.x.1948, *Barbosa* 2436 (K, LISC, LMA). T: Moatize Dist., Revúboè R., near Tete, fl.& fr. x.1883, *Kirk* s.n. (K). MS: Mossurize Dist., Zinyumbo Hills, 450 m, fl. 23.xi.1906, *Swynnerton* 961 (K). GI: Zavala Dist., Deropa stream, fl.& fr. 31.viii.1975, *Moura et al.* 345 (LISC). M: Namaacha Dist., Namaacha Falls, fr. 22.ii.1955, *Exell, Mendonça & Wild* 544 (BM, LISC).

Also in tropical E Africa, Congo, Angola and throughout southern Africa. Often dominant in rivers, swamps, lagoons, dams and along watercourses; 0–1350 m (c.2000 m outside the Flora area).

Conservation notes: Widespread and common species over much of Africa.

According to Galen Smith (1987) and Cook (2004), it is quite likely that *T. capensis* has arisen through hybridization between *T. latifolia* and *T. domingensis*, followed perhaps by polyploidy. Some plants are female sterile with small withered ovaries.

RESTIONACEAE

by H.P. Linder

Evergreen, rush-like, generally dioecious plants, tangled, mat-forming, caespitose or rhizomatous. Culms green, photosynthetic, simple or branched, terete or compressed, smooth, striate or sulcate, epidermis smooth, tuberculate or pimpled, rarely hirsute. Prophylls glabrous or villous. Leaves reduced to sheaths; sheaths falling early or persistent, split to the base, margins various, often membranous or translucent, apex often with large translucent lobes, usually mucronate or awned, awns occasionally large and leaf-like. Inflorescence often sexually dimorphic, racemose or paniculate, bracteate. Flowers usually aggregated into spikelets, often with sterile bracts, sometimes racemose, never solitary. Perianth of 6 tepals arranged in 2 whorls, rarely the number reduced; tepals similar, or the two whorls differentiated, or the outer lateral tepals conduplicate, often deeply keeled or villous along the keel, cartilaginous, bony, papery or membranous. Androecium of 3 stamens opposite the inner whorl of tepals; filaments usually free, anthers 1-locular, versatile. Ovary superior, 1–3-locular; ovules solitary on each locule, pendulous; styles 1–3, usually free, if 2 occasionally fused below. Fruit a capsule or nut; seeds variously ornamented.

A very distinct family of about 50 genera and 450 species known from southern Africa, Australia and New Zealand, with a few species in central and south-central Africa, Madagascar, Malaysia, Indo-China and southern Chile. The main African centre of diversity is in the Western Cape Province of South Africa; only 1 genus in the Flora area.

PLATYCAULOS H.P. Linder

Platycaulos H.P. Linder in Bothalia **15**: 64 (1984). —Linder & Hardy in Bothalia **40**: 6–8 (2010)

Plants caespitose or mat-forming, rarely rhizomatous. Culms ± compressed, branching, prophylls usually villous, surface smooth, pimpled or tuberculate. Sheaths tightly convoluted, persistent, the margins various, apex awned or mucronate, awn never leaf-like. Male inflorescence racemose or paniculate, flowers aggregated into spikelets, subtended by spathes or spathellae; floral bracts imbricate, cartilaginous to bony, usually at least as tall as the flowers; flowers very shortly pedicillate, perianth papery to subcartilaginous; tepals subequal, but with outer lateral ones usually conduplicate, glabrous or villous; anthers unilocular, exserted from perianth at anthesis; pistillode minute. Female inflorescence often different from the male with fewer and larger spikelets; spikelets often with sterile bracts and a solitary flower; perianth bony to cartilaginous, occasionally the inner whorl papery, usually the outer lateral tepals folded together, keels glabrous or villous, usually slightly taller than inner whorl; staminodes always present; ovary 1–2(3)-locular; styles 3, free. Fruit a capsule, seeds variously ornamented.

A genus of 12 species endemic to Africa and Madagascar, most diversified in the Western Cape Province of South Africa.

A revision of the African members of the family (Linder in Bothalia **15**: 11–76, 1984) has split the genus *Restio*, recognising largely three groups: *Restio sensu stricto*, *Ischyrolepis* and *Platycaulos*. Molecular phylogenetic work (Hardy *et al.* in *Inter. J. Pl. Sci.* **169**: 377–390, 2008) showed that *Ischyrolepis* was nested in *Restio*, but that *Platycaulos* was indeed distinct. Furthermore, it showed that the tropical African members of *Restio* should be grouped with *Platycaulos*. Careful morphological investigation revealed that these taxa have the critical characters of *Platycaulos*, i.e. a somewhat compressed culm and dense green sheaths. Consequently the species from the Flora area have to be placed under *Platycaulos* rather than *Restio*.

1. Bracts of female flowers at least twice as long as flowers · · · · · · · **1.** *quartziticola*
 – Bracts of female flowers less than twice as long as flowers· · · · · · · · · · · · · · · 2
2. Plants tufted; flowering December–January · · · · · · · · · · · · · · · · · **3.** *mlanjiensis*
 – Plants spreading; flowering April–May· **2.** *mahonii*

1. **Platycaulos quartziticola** (H.P. Linder) H.P. Linder & C.R. Hardy in Bothalia **40**: 8 (2010). Type: Zimbabwe, Chimanimani Mts, ♀ fl. 23.iv.1957, *Whellan* 1252 (K holotype, BOL, BR, LISC, SRGH).

 Restio quartziticola H.P. Linder in Kew Bull. **41**: 103 (1986).

Plants ± caespitose, to 100 cm tall. Culms apically slightly compressed, branching, 0.5–2 mm in diameter at base, surface finely wrinkled to obscurely tuberculate, rarely pimpled. Sheaths tightly convoluted, 5–13 mm long, acute to obtuse, awn 1–3 mm long, somewhat swollen towards apex; upper margins submembranous, dark brown with a narrow translucent edge. Male inflorescence racemose with 2–10 spikelets; spathes like sheaths, 5–10 mm long, falling at flowering; spikelets 4–15 mm long, with 1–7 flowers; bracts papery, acute, 3–7 × 1–2 mm, slightly taller to twice as tall as flowers; flowers shortly pedicellate; perianth 2.5–3.5 mm long, outer lateral tepals conduplicate, glabrous or rarely scabrid or hispid on the keel, apices finely acute, often recurved, subcartilaginous, inner whorl slightly shorter than outer, flat, papery; anthers 1.5–2 mm long; pistillode minute, 3-lobed. Female inflorescence with 1–4 racemose spikelets; spathes 7–13 mm long, falling at flowering time; spikelets 8–15 mm long, 3–8 flowered; bracts papery, acute, more than twice as long as flowers, 6–12 × 1–2 mm; flowers shortly pedicellate, 3–4 mm long, perianth lobes unequal, the outer whorl cartilaginous with the lateral folded together, keels glabrous or finely scabrid, the middle tepal flat, marginally shorter; inner whorl papery to subpapery, flat, acute, 3–3.5 × 1 mm; staminodes 3, c.1 mm long; ovary bilocular; styles 3. Fruit a 2-locular capsule; seed 1.5 mm long, ellipsoid, surface rugose with a finely reticulate pattern.

Zimbabwe. E: Chimanimani Mts, Point 71, ♀ fl. 16.iii.1957, *Phipps* 646a (BOL, BR, K, LISC). Chimanimani Mts, Bundi, ♂ fl. 17.iv.1957, *Goodier & Phipps* 1 (BOL, BR, K, LISC). **Mozambique.** MS: Sussundenga Dist., Mevumozi R. tributary, ♀ fl. 6.v.1965, *Whellan* 2231 and ♂ fl. 6.v.1965, *Whellan* 2232 (K, LISC).

Not known elsewhere. Appears to be restricted to quartzite substrates, and usually prefers moist or boggy habitats or rocky ledges; above 900 m.

Conservation notes: Endemic to the Chimanimani Mts. Although the species is very restricted in its global distribution, it appears to be common within this area and cannot be regarded as Vulnerable.

Platycaulos quartziticola was previously included in *Restio mahonii* from Malawi.

2. **Platycaulos mahonii** (N.E. Br.) H.P. Linder & C.R. Hardy in Bothalia **40**: 8 (2010). Type: Malawi, Zomba, ♂ fl. 5.xii.1898, *Mahon* s.n. (K holotype, B, BOL). FIGURE 13.4.**13**.

 Hypolaena mahonii N.E. Br. in F.T.A. **8**: 265 (1901). —Pillans in Trans. Roy. Soc. S. Afr. **16**: 398 (1928).

Restio madagascariensis Cherm. in Bull. Soc. Bot. Fr. **69**: 318 (1922). —Humbert in Fl. Madag. **34**: 3 (1946). Type: Madagascar, Mt Ibity, 1800–2300 m, ♂, *Perrier de la Bâthie* 2735 (P lectotype).

Restio mahonii (N.E. Br.) Pillans in Trans. Roy. Soc. S. Afr. **30**: 255 (1945). —Linder in Kew Bull. **41**: 100 (1986). —Beentje in F.T.E.A., Restionaceae: 1 (2005).

Subsp. **mahonii**.

Plants clumped, often spreading to mat-forming, to 100 cm tall. Culms terete, 1–2(3) mm in diameter, terminal branchlets occasionally ± compressed, branching, surface wrinkled and generally finely pimpled. Sheaths closely convoluted, 8–15 mm long, obtuse to acute, awn 0.3–5 mm long, body olive-green, margins distinct, reddish brown. Male inflorescence with 1–8 racemosely arranged, 5–10 mm long, lax, several-flowered, spikelets; spathes 5–8 mm long; bracts cartilaginous to leathery, brown with pale margins, ovate-elliptical, 3–5 mm long, generally as long as perianth; flowers shortly pedicellate, 3–4 mm long; tepals papery to cartilaginous, acute apices dark brown; outer lateral ones folded double with keel glabrous to rarely pilose; odd outer tepal flat, inner tepals shorter and paler than outer whorl; anthers 1.7–2.4 mm long; pistillode 3-lobed, minute. Female inflorescence with 1–4, compact, 2–8-flowered, 6–12 mm long spikelets; spathes ²/₃ of spikelet length, persistent; bracts marginally taller than flowers, 4–6 mm long, leathery, ovate, acute; flowers shortly pedicellate; perianth glabrous or villous on keels, outer lateral tepals keeled, odd outer tepal and inner tepals strap-shaped, obtuse; outer tepals 2.8–4 mm long, inner tepals slightly shorter; staminodes c.1 mm long; ovary 2-locular, styles 3. Fruit a 2-locular capsule. Seeds pale brown, ellipsoid, very finely sculptured.

Malawi. S: Mt Mulanje, Lichenya Plateau, ♀ fl. 7.vi.1962, *Robinson* 5291 (K, LISC); & ♂ fl. 7.vi.1962, *Robinson* 5292 (K, LISC).

Also known from Congo, Tanzania and Madagascar. Generally associated with the high montane flora in peaty, swampy or wet places, or in rock-cracks; above 1500 m.

Conservation notes: Restricted distribution, but not under particular threat; probably Near Threatened.

In Malawi this subspecies shows little variation, but over the rest of its range the sheaths are generally loosely convoluted, occasionally even spreading, the sheath margins may be translucent, and the bracts and flowers larger.

A second subspecies, subsp. *humbertii* (Cherm.) H.P. Linder, occurs in Madagascar and is distinguished by the capitate inflorescence and usually bisexual flowers.

3. **Platycaulos mlanjiensis** (H.P. Linder) H.P. Linder & C.R. Hardy in Bothalia **40**: 8 (2010). Type: Malawi, Mt Mulanje, Sombani, ♀ fl. 10.xii.1991, *Verboom, Pauw & Hooks* 1 (BOL holotype, K, MAL, MO, NSW, PRE, S, SRGH, WAG, Z).

Restio mlanjiensis H.P. Linder in Kew Bull. **50**: 623 (1995).

Plants caespitose, 40–100 cm tall. Culms terete, 0.8–2 mm in diameter, somewhat compressed, finely pimpled, olive-coloured. Sheaths closely convoluted, 10–16 mm long, obtuse to acute, mucro 0.8–1.2 mm long, body olive-green, margins narrowly membranous and reddish. Male inflorescence with 4–7 racemosely arranged, 5–8 mm long, 4–6 flowered spikelets; spathes persistent, c. ²/₃ of spikelet length; bracts leathery, slightly taller than flowers, 5–6 mm long, ovate, acute; flowers shortly pedicellate, 3–4 mm long; tepals papery to cartilaginous, glabrous, outer lateral tepals folded double; anthers 1.5 mm long, exserted from flowers; pistillode minute. Female inflorescence with 2–5, compact, 6–8-flowered, 8–12 mm

---

Fig. 13.4.**13**. PLATYCAULOS MAHONII. 1, habit, female plant (× ½); 2, male flower spikelet (× 3); 3, male flower (× 6); 4, male flower, anthers and pistillode rudiments (× 6); 5, female flower spikelet (× 3); 6, female flower (× 6); 7, fruit (× 6). All from *Robinson* 5291 & 5292. Drawn by Juliet Williamson. From Flora of Tropical East Africa.

long spikelets; spathes c. $^2/_3$ of spikelet length, persistent; bracts slightly taller than flowers, 4–5 mm long, leathery, ovate, acute; flowers shortly pedicellate, perianth glabrous; outer lateral tepals keeled, odd outer tepal and inner tepals strap-shaped, obtuse; outer tepals 2.5–3.5 mm long, inner tepals 2.4–3.2 mm long; staminodes present; ovary 2-locular, styles 3. Fruit a 2-locular capsule. Seeds 1— 0.8 mm, black, with uneven surface.

Malawi. S: Mt Mulanje, Sombani, ♀ fl. 10.xii.1991, *Verboom, Pauw & Hooks* 1 (BOL, K, MAL, MO, PRE, SRGH, WAG, Z).

Only known from the Sombani Plateau on Mt Mulanje, where it grows in grassland; 1700–2200 m.

Conservation notes: Endemic to one locality on Mt Mulanje where it is very rare; could be sensitive to changes in fire regime. Possibly Endangered.

Platycaulos mlanjiensis is very similar to *P. mahonii*, but differs primarily in its growth-form. *P. mahonii* is a spreading plant, with the basal parts of the culms decumbent and often rooting, hence it tends to be mat-forming. *P. mlanjiensis*, by contrast, is a tufted plant without decumbent culms. This suggests that *P. mlanjiensis* is killed by fire and that populations are re-established from seedlings, while *P. mahonii* survives fire and does not rely on seedling establishment. There might also be differences in the seed colour, but not enough specimens have been seen to corroborate this.

FLAGELLARIACEAE

by H.P. Linder

Perennial herbs, climbing by means of leaf-tip tendrils. Leaves alternate, glabrous; sheath entire or split to the base, lamina flat with a long tendril-like apex, abruptly narrowed at base, venation parallel. Inflorescence a panicle. Flowers sessile or subsessile, aggregated into glomerules or small racemes, usually hermaphrodite, small, bracteate, bisexual, regular; perianth of 6 subequal, thin, persistent tepals arranged in 2 whorls. Stamens 6, hypogynous or attached to base of tepals; filaments short, free; anthers 2-thecous, opening by longitudinal slits. Ovary superior, 3-locular, ovules pendulous, 1 per locule; styles 3, free. Fruit a 1–3 seeded berry.

A small, monogeneric family, formerly including *Joinvillea* which is now separated in the Joinvilleaceae.

FLAGELLARIA L.

Flagellaria L., Sp. Pl.: 333 (1753).

Description as for the family.

A genus of 3–4 species, widespread in the Old World Tropics, with one species in Africa and another in Madagascar.

Flagellaria guineensis Schumach., Beskr. Guin. Pl.: 181 (1827). —Brown in Fl. Cap. **7**: 16 (1897). —Rendle, Cat. Afr. Pl. Welw. **2**: 81 (1899). —Brown in F.T.A. **8**: 90 (1901). —Hepper in F.W.T.A. ed.2, **3**: 51 (1968). —Napper in F.T.E.A., Flagellariaceae: 1 (1971). Type: Ghana (Guinea), *Thonning* s.n. (C holotype). FIGURE 13.4.**14**.

Flagellaria indica L. var. *guineensis* (Schumach.) Engl., Pflanzenw. Afrikas **2**: 257, fig.174 (1908). —Marloth, Fl. S. Afr. **4**: 56, fig.11 (1915).

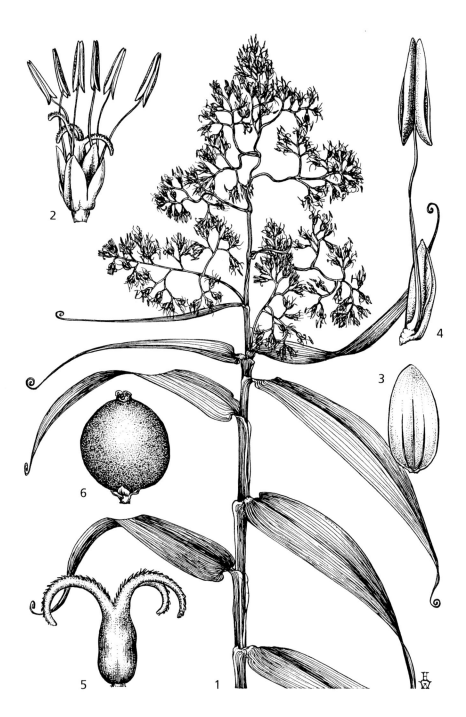

Fig. 13.4.**14**. FLAGELLARIA GUINEENSIS. 1, shoot with leaf-sheath and inflorescence (× ²/₃); 2, flower (× 5); 3, perianth segment (× 10); 4, stamen with perianth segment (× 10); 5, gynoecium (× 10); 6, fruit (× 4). 1 from *Tanner* 3736, 2–6 from *Faulkner* 1609. Drawn by Heather Wood. From Flora of Tropical East Africa.

Perennial herb climbing by means of leaf-tip tendrils, culms to 5(10) m, 0.5–1.5 cm in diameter, glabrous. Leaves alternating; sheaths imbricate, split to the base, with a narrow papery margin, subauricular at apex; lamina very narrowly lanceolate, 7–25 × 1–2.8 cm, acute, with apex continued into a tendril, glabrous, shining, veins parallel. Inflorescence a pyramidal panicle, 3–12 cm long; flowers subsessile in short racemes; bracts 0.5–1 mm long, oblate, persistent. Perianth segments 6, oblong, rounded, 2–3 mm long, papery, white or cream, arranged in 2 whorls with inner whorl somewhat taller than outer. Stamens 6, filaments free, anthers 1.5–2 mm long, exserted from perianth at anthesis. Ovary 3-locular, styles 3, free, maturing after anthers have fallen; ovules pendulous, solitary in each locule. Fruits clustered, a berry, green turning red or orange, 3–5 mm in diameter.

Mozambique. N: Pemba Dist., Nangororo, rio Ridi, fl. 28.x.1959, *Gomes e Sousa* 4494 (COI, K). Z: Lugela Dist., Namagoa, fl. ix-x.1944, *Faulkner* 25 (BM, COI, K). MS: Machanga Dist., 40 km from Nova Mambone, on road to Buzi, fl.& fr. 2.ix.1942, *Mendonça* 116 (BR, COI, K, LISC, LMA, LMU, PRE). GI: Guijá Dist., Aldeia da Barragem, L bank of rio Limpopo, fr. 16.xi.1957, *Barbosa & Lemos* 8148 (COI, K, LISC). M: Matutuine Dist., between Santaca and Frazão, fl. 24.ix.1948, *Gomes e Sousa* 3850 (COI, K).

Within the Flora area only known from Mozambique. Also in Cameroon, Congo, Gabon, Togo, Benin, Nigeria, Ivory Coast, Ghana, Angola, and in Somalia, Kenya, Tanzania and South Africa. Although the species is mostly coastal, it is also widespread in the Congo Basin. In moist to wet, tropical, coastal, swampy or riverine forests; below 600 m.

Conservation notes: Widespread species, locally common; not threatened.

Generally plants occur near the forest edge, along rivers, clearings, roads and where there has been vegetation disturbance. The species is a powerful climber able to reach the forest canopy, where it usually flowers. The liana-like stems are used locally for basket-weaving.

Napper in F.T.E.A. recorded a single collection of *F. indica* from Tanzania, and it is also listed in da Silva, Izidine & Amude (Prelim. Checklist Pl. Moz., 2004) from Niassa. It has not been possible to trace any specimens of this species from continental Africa, so the presence of this second species cannot be confirmed. *F. indica* can be readily separated from *F. guineensis* by the entire leaf-sheaths and the flowers which are congested in short racemes. *Flagellaria indica* is recorded from Madagascar and the northern and eastern borders of the Indian Ocean, reaching E Australia, while *F. guineensis* is endemic to Africa. I have not been able to confirm an overlap in their distributions.

JUNCACEAE

by F.M. Crawford

Annual or perennial herbs, usually rhizomatous. Roots fibrous. Stems erect, rarely procumbent, cylindrical, rarely compressed, naked or leaf bearing. Leaves basal or cauline (arising from stem), cylindrical, pungent, rigid or soft, rarely flat or channelled, sometimes divided by partitions (septate), margins glabrous or hairy; basal leaves sometimes reduced to scale-like leaves (cataphylls), basal part of leaf enveloping stem (sheath) open or closed, sometimes with long hairs or auriculate. Flowers few to many, aggregated into a terminal or pseudo-lateral panicle (bract resembling continuation of the stem); arranged in heads or capitula. Inflorescence bracts leafy, cylindrical or filiform, bracteoles present, 1–2 or absent. Flowers small, regular, usually bisexual, rarely unisexual; perianth of 6 glume-like segments in 2 whorls of 3, ± equal, free. Stamens 3 or 6, shorter than perianth; filaments filiform to triangular; anthers basifixed, bithecous, dehiscing longitudinally. Ovary superior, carpels 3, forming a 1- or 3-locular ovary with 3 to many ovules; style 1, 3-branched. Fruit a longitudinally dehiscing capsule. Seeds 3 to many, small, ovoid, obovoid or globose, usually with basal or apical appendages.

A family comprising 7 genera and c.450 species, most with a worldwide distribution and most common in temperate regions. Only 2 genera in Africa, with the remaining 5 in South America and New Zealand; 1 genus in the Flora area.

Juncaceae is closely related to Cyperaceae; the two may be confused but can be easily separated by the fruit and flower – Juncaceae has a capsule while Cyperaceae has a nutlet. A regular flower with 6 glume-like perianth segments is characteristic of both families, however it is much reduced or absent in Cyperaceae.

Juncaceae is closely related to Thurniaceae (= Prionaceae) and Mayacaceae, both with disjunct distributions.

JUNCUS L.

Juncus L., Sp. Pl.: 325 (1753). —Kirschner *et al.*, Juncaceae 2 in Orchard *et al.*, Species Plantarum: Fl. World **7**: 16 (2002).

Annual or perennial herbs, glabrous. Roots thin and numerous or thick and woody, sometimes with dense covering of root hairs. Plant base rhizomatous, tufted, caespitose. Stems erect or procumbent, cylindrical to compressed. Leaves basal or cauline, rigid, pungent, filiform, cylindrical or flat, sometimes channelled or septate; basal leaves often reduced to cataphylls. Sheaths at base of leaves open, rarely closed, often with distinct auricles at connection to the blade. Main inflorescence bracts leaf-like or reduced. Inflorescence a cluster of compound terminal or pseudo-lateral panicles, sometimes reduced to 1- or 2-flowered capitula subtended by a bract or 2 bracteoles. Flowers bisexual, rarely unisexual. Perianth of 6 free, glume-like segments in 2 whorls of 3. Tepals glumaceous, equal or unequal, ovate or lanceolate, obtuse to acute, green to brown. Stamens 3 or 6, filaments filiform or flattened, anthers linear to oblong. Ovary superior; style present or absent; stigmas 3. Capsules 1- or 3-locular. Seeds numerous and small, smooth or reticulate, sometimes with basal and/or apical appendages.

A large genus with c.220 species widely distributed in temperate regions; also found in the artic and at high altitudes in the tropics. Common in temporary moist depressions, marshes, saltpans and along riverbanks.

Some species of *Juncus* are ecologically important and may be locally dominant in swampy areas. *Juncus* may be wind pollinated (anemophily) or self-pollinated (autogamy), and some species may spread by vegetative buds developing in the inflorescences (vivipary). Seeds are dispersed in water and by animals.

There have been attempts to separate *Juncus* into a number of genera, none of which have been successfully maintained. The taxonomy used here follows Kirschner *et al.* (2002).

The name *Juncus* is derived from the Latin word 'jungere' meaning to join, referring to its use in ancient times for producing plaited mats or chair seats. It was also used for paper making, while the pith of some species has been used as wicks.

1. Leaves cylindrical, rigid, pungent or soft, sometimes septate; cataphylls well developed · · · · · 2
 – Leaves flat or channeled, not septate; cataphylls absent · · · · · 7
2. Leaves not septate; inflorescence with few to many flowers (2–10) in a loose capitulum · · · · · 3
 – Leaves septate; inflorescence with many flowers (4–30) in a compact capitulum · · · · · 5
3. Tepals ovate; each flower subtended by a single bract; capsule acuminate, brown to dark reddish brown, shiny · · · · · 4
 – Tepals linear-lanceolate; each flower subtended by a bract and 2 bracteoles; capsule rounded at apex, yellow to light brown · · · · · **8.** *effusus*

4. Tepals uniformly coloured (green to straw-yellow) without a darkened midrib; capsule mostly exserted from perianth; seeds fusiform 0.5–0.7 mm long with long white appendages at each end, 0.8–1.2 mm long; in brackish salt-pan areas · **1.** *rigidus*
- Tepals with a darkened midrib; capsule ± equal to perianth in length; seeds ellipsoid, 0.6–0.8 mm long, blunt to pointed at one end, with a distinct white appendage at the other, 0.8–0.9 mm long; coastal species, rarely inland · **2.** *kraussii*
5. Flowering stem with one long leaf; capsule trilocular · · · · · · · · · · · **5.** *punctorius*
- Flowering stem with 2–5 leaves; capsule unilocular · · · · · · · · · · · · · · · · · · · 6
6. Capsule exserted from perianth · **7.** *exsertus*
- Capsule equal in length or shorter than perianth · · · · · · · · · · · · · · **6.** *oxycarpus*
7. Leaves flat, greater than 3 mm wide · **3.** *lomatophyllus*
- Leaves channeled, smaller than or equal to 3 mm wide · · · · · · · · · **4.** *dregeanus*

1. **Juncus rigidus** Desf., Fl. Atlant. **1**: 312 (1800). —Obermeyer in F.S.A. 4(2): 79 (1985). —Snogerup in Willdenowia **23**: 54 (1993). —Clarke & Klaassen, Water Pl. Namibia: 82 (2001). —Cook, Aquat. Wetl. Pl. Sthn. Africa: 155 (2004). Type: NW Africa, 'in sand by sea shore', *Desfontaines* s.n. (P lectotype).

   Juncus maritimus Lam. var. *rigidus* (Desf.) Rouy, Fl. France **13**: 230 (1912).

   Juncus arabicus (Asch. & Buchenau) Adamson in J. Linn. Soc., Bot. **50**: 10 (1935). Type: Egypt, W Sinai, 17.vii.1835, *Schimper* 495 (W lectotype, CAS, E, FI, M), lectotypified by Snogerup (1993).

Perennial herb, 60–120 cm tall. Rhizome compact, creeping, woody; roots thick, covered in dense root hairs. Stems terete, hard. Leaves basal, cataphylls present, sheaths auriculate; blades terete, not septate. Inflorescence a lax pseudo-lateral panicle, subtended by a bract, a continuation of the stem; flowers 60–150, each capitula with 2–6 flowers; floral bract solitary (bracteole absent). Tepals narrowly ovate with membranous margins, straw-coloured (may be slightly darker); outer tepals equal or slightly longer (2.9–3.5 mm) than inner (2.4–2.9 mm). Stamens 6, shorter than tepals. Capsule trilocular, narrowly cylindrical or ovoid, 2.5–3.5 mm long, apiculate, as long or slightly longer than tepals. Seeds numerous, fusiform, 0.8–1.2 mm long including distinct white appendages at each end.

Botswana. N: Makgadikgadi Pans, c.100 m from edge of Boteti R., 21°03.25'S 24°58.15'E, 29.ii.1980, *P.A. Smith* 3117 (K, SRGH). **Zambia**. B: Senanga Dist., Angola border, Mashi River fringe, 28.ix.1964, *Verboom* 1756 (K). C: Mumbwa, 4 miles S of Namabeshi, Narubanda along edge of Kafue floodplain, 9.ix.64, *van Rensburg* KBS 2959 (K). S: Choma Dist., Muckle Neuk, 19 km N of Choma, 10.x.1954, *Robinson* 902 (K). **Zimbabwe**. N: Binga Dist., Binga, Chikwatala Hot Springs, 7.xi.1958, *Phipps* 1391 (K, PRE, SRGH). C: Kadoma Dist., 15.xi.1970, *Wild* 7798 (K, PRE, SRGH). S: Chiredzi Dist., Chiredzi R., Nandi Farm, 2.ix.1969, *Jaklitsch* 197.280 (K, LISC, PRE, SRGH).

Also in Angola, Caprivi Strip, Lesotho, Malawi, Swaziland, the Mediterranean region, Arabian Peninsula and Asia; a widespread halophyte in Southern Africa, forming large colonies. In sandy and saline localities, coastal or inland, and in drier areas next to wells and seasonally wet depressions; sea level to 1300 m.

Conservation notes: Widespread species; not threatened.

Juncus rigidus is closely related to *J. kraussii* but the flowers are more loosely arranged, the narrow capsule is distinctly exserted beyond the perianth and the seeds have long appendages at either end.

2. **Juncus kraussii** Hochst. in Flora **28**: 342 (1845). —Obermeyer in F.S.A. **4**(2): 78 (1985). —Snogerup in Willdenowia **23**: 57 (1993). Type: South Africa, George Dist., Notsinakama R, i.1839, *Krauss* s.n. (G lectotype, M), lectotypified by Snogerup (1993).

Perennial herb, 50–100 cm tall. Rhizome robust, compact, creeping; roots thick covered in dense root hairs. Stems erect, terete, rigid. Leaves basal, cataphylls present, sheaths auriculate; blades terete, pungent, not septate. Inflorescence a lax pseudo-lateral panicle, with subtending bract, a continuation of the stem; each capitula with 2–6 flowers; floral bract solitary. Tepals acuminate, yellow with red margins, becoming harder and colourless with age, outer tepals longer (2.7–3.2 mm) than inner (2.3–2.7 mm). Stamens 6. Capsule apiculate, woody, shiny, red to yellow, enclosed in or slightly longer than perianth. Seeds ± 0.7 mm long, ellipsoid, proximal end blunt, distal end pointed, with distinct white appendage, reticulate.

Subsp. **kraussii**. —Kirschner *et al.*, Juncaceae 2 in Orchard *et al.*, Species Plantarum: Fl. World **7**: 20 (2002). —Cook, Aquat. Wetl. Pl. Sthn. Africa: 154 (2004).

Juncus caffer Bertol. in Mem. Reale Accad. Sci. Ist. Bologna **3**: 253 (1851). Type: Mozambique, Inhambane, 6.xii.1848, *Fornasini* s.n. (BOLO holotype).

Juncus fasciculiflorus Adamson in J. Linn. Soc., Bot. **50**: 8 (1935). Type: South Africa, Ceres, Zwartruggens, *Adamson* 123 (BOL holotype).

Juncus kraussii var. *parviflorus* Adamson in J. Linn. Soc., Bot. **50**: 8 (1935). Type: South Africa, Buffalo R., 24 km from E London, 18.i.1930, *Adamson* 179 (BOL lectotype).

Juncus kraussii var. *effusus* Adamson in J. Linn. Soc., Bot. **50**: 8 (1935). Type: South Africa, Port Alfred, 24.xii.1898, *Galpin* 2946 (K lectotype).

Inflorescence lax, rarely congested; most flowers unisexual and functional.

Mozambique. GI: Inhassoro Dist., Bazaruto Is., 28.x.1958, *Mogg* 28700 (K, LISC). M: Marracuene, 22 xi.1979, *de Koning* 7652 (K).

Also in Lesotho and South Africa. In strongly saline marshes, along rivers and rarely along coastal sands; sea-level to 1700 m.

Conservation notes: Widespread taxon; not threatened.

Three disjunct subspecies are recognised: subsp. *kraussii* from Mozambique to South Africa, subsp. *australiensis* (Buchenau) Snogerup from Australia, Tasmania and New Zealand and subsp. *austerus* (Buchenau) Snogerup from temperate South America.

The stems are a popular material for weaving, particularly for sleeping mats, especially in parts of KwaZulu-Natal.

3. **Juncus lomatophyllus** Spreng., Neue Entdeck. Pflanzenk. **2**: 108 (1821). —Baker in Fl. Cap. **7**: 27 (1897); in F.T.A. **8**: 94 (1901). —Obermeyer in F.S.A. **4**(2): 82 (1985). —Cook, Aquat. Wetl. Pl. Sthn. Africa: 154 (2004). Type: South Africa, Cape Peninsula, *Bergius* s.n. (B† holotype). Neotype: South Africa, Cape of Good Hope, n.d., *Cole* s.n. (K neotype, designated here).

Juncus cymosus Lam., Encycl. **3**: 267 (1789). Type: South Africa, Cape of Good Hope, *Sonnerat* s.n. (P lectotype), lectotypified by Obermeyer (1985).

Juncus capensis Thunb. var. *latifolius* E. Mey., Syn. Junc.: 48 (1822). Type: no locality, no collector (S lectotype, TCD), lectotypified by Kirscher *et al.* (2002).

Juncus lomatophyllus var. *lutescens* Buchenau in Abh. Naturwiss. Vereine Bremen **4**: 466 (1875). Type: South Africa, Dutoitskloof, *Drège a* (W syntype).

Juncus lomatophyllus var. *aristatus* Buchenau in Abh. Naturwiss. Vereine Bremen **4**: 466 (1875). Type: South Africa, Dutoitskloof, *Drège f* (W lectotype), lectotypified by Obermeyer (1985).

Juncus lomatophyllus var. *congestus* Adamson in J. Linn. Soc., Bot. **50**: 18 (1935). Type: South Africa, Swartberg Pass, 4.ii.1930, *Adamson* 181 (BOL lectotype), lectotypified by Obermeyer (1985).

Juncus viridifolius Adamson in J. Linn. Soc., Bot. **50**: 20 (1935). Type: South Africa, Cape, Kirstenbosch, Table Mountain, Skeleton Ravine, 11.x.1931, *Adamson* 247 (BOL holotype).

Perennial herb, 15–85 cm tall. Rhizome densely creeping, stolons often present; roots thin. Stems rigid, terete, with longitudinal grooves. Leaves basal, cataphylls absent; blades flat, broadly linear, up to 15 mm wide, blue-green with an occasional red tinge, not septate. Inflorescence a lax panicle, appearing terminal due to subtending leaf-like bract; capitula 10–50, semi-globose, each with 2–12 flowers; bracteoles absent. Tepals: outer 2.9–3.5 mm long, dark brown, sometimes black, with aristate apex; inner tepals shorter, 2.5–3 mm long, more acute, margins membranous. Stamens 6. Capsule ovoid, trilocular, 2.1–2.6 mm long, much shorter than tepals. Seeds numerous, 0.4–0.6 mm long, oblong or ovoid, apiculate, apex white, brown, surface faintly reticulate.

Zimbabwe. E: Nyanga Dist., Nyanga Nat. Park, Mare Dam, 6.i.1972, *Gibbs Russell* 1203 (K, PRE, SRGH). **Mozambique**. MS: Gorongosa Mt, Gogogo summit area, x.1971, *Tinley* 2216 (K, LISC, SRGH). M: Marracuene Dist., near Bobole, Incomati valley, Chinese farm, 2.x.1957, *Barbosa & Lemos* 7919 (K, LISC, LMA).

Also in Lesotho, South Africa and Swaziland. In permanently wet areas, such as streams, edges of ponds and reservoirs; sea-level to 1700 m.

Conservation notes: Widespread in the Flora area; not threatened.

Very variable in growth form due to altitude and climate.

Juncus lomatophyllus is closely related to *J. engleri* Buchenau, which is only known from Tanzania.

4. **Juncus dregeanus** Kunth, Enum. Pl. **3**: 344 (1841). —Baker in Fl. Cap. **7**: 25 (1897). —Obermeyer in F.S.A. **4**(2): 84 (1985). —Cook, Aquat. Wetl. Pl. Sthn. Africa: 153 (2004). Type: South Africa, "between Cape Colony and Port Natal", *Drège* 4387 (B† holotype). Neotype: South Africa, KwaZulu-Natal, Port Shepstone, 30.ix.1965, *Ward* 5209 (K neotype, PRE), designated here.

Juncus dregeanus var. *genuinus* Buchenau in Abh. Naturwiss. Vereine Bremen **4**: 463 (1875). Type as above.

Perennial herb, 10–50 cm tall. Rhizome small; roots thin, numerous. Stems erect. Leaves basal, cataphylls absent, sheaths initially closed, not auriculate; blades linear, 0.8–3 mm wide, channelled with inrolled margins, numerous, not septate. Inflorescence a terminal panicle, bract leaf-like; capitula 3–8, young capitula semi-globose becoming globose with age, pedunculate, each with 5–30 flowers; floral bract solitary, bracteoles absent. Tepals yellow-brown with a darkened midrib and tip; outer tepals longer, aristate, 2.6–3.3 mm long; inner tepals shorter, acute, 2.5–3 mm long with hyaline margin. Stamens 3, rarely 6. Stigmas spreading. Capsule subglobose, acuminate, ± 2.5 mm long, red-brown to yellow almost black at apex, apiculate, shorter than perianth. Seeds 0.35–0.55 mm long, reticulate with white appendages.

Subsp. **bachitii** (Steud.) Hedberg in Symb. Bot. Upsal. **15**: 61 (1957). —Carter in F.T.E.A., Juncaceae: 4 (1966). —Haines & Lye, Sedges Rushes E Africa: 32 (1983). Type: Ethiopia, Tigrey, Mt Bachit, *Schimper* 114 (P holotype, PRE, W). FIGURE 13.4.**15**.

Juncus bachitii Steud., Syn. Pl. Glum. **2**: 305 (1855).

Tepals almost entirely dark brown to black-brown apart from pale marginal border. Stamens 6.

Zambia. E: Chama Dist., Nyika Plateau, 2.i.59, *Robinson* 3008 (K, PRE). **Zimbabwe**. E: Mutare Dist., Lake Alexander, 7.i.1972, *Gibbs Russell* 1251 (K, PRE). **Malawi**. N: Rumphi Dist., Nyika Plateau, Lake Kaulime, 4.i.1959, *Robinson* 3035 (K, PRE). S: Mt Mulanje, Lichenya Plateau, 29.xii.1988, *Chapman et al.* 9458 (K, MO). **Mozambique**. MS: Sussundenga Dist., Chimanimani Mts, path between Skeleton Pass and Namadima, 30.xii.1959, *Goodier & Phipps* 357 (K, PRE, SRGH).

Fig. 13.4.**15**. JUNCUS DREGEANUS subsp. BACHITII. 1, habit (× ²/₃); 2, flower and bract (× 6); 3, style and stigmas (× 12); 4, seed (× 60). JUNCUS OXYCARPUS. 5, habit (× 1); 6, part of inflorescence (× ¹/₂); 7, flower in fruit (× ¹/₁₀). JUNCUS RIGIDUS. 8, fusiform seed (bar = 1 mm). JUNCUS KRAUSSII subsp. KRAUSSII. 9, seeds (bar = 0.5 mm). Figs. 1–3 from *Fries* 1189, 4 from *Brenan & Greenway* 8218, drawn by Margaret Stones, from Flora of Tropical East Africa; figs. 5–7 from *Dieterlen* 767b, drawn by R. Holcroft, from Flora of Southern Africa, reproduced with permission; fig. 8 from *Zohary & Yehoudai* 721; fig. 9 from *F. Müller* s.n., figs. 8–9 drawn by B. Johnsen, from Species Plantarum.

Also in Kenya, Tanzania and Uganda. In areas of swampy ground and shallow water, edges of streams and marshes, mainly in montane forests, rarely in wet alpine grasslands; above 2000 m.

Conservation notes: Widespread species; not threatened.

Subsp. *dregeanus* occurs in South Africa, Lesotho and Swaziland. It differs from subsp. *bachitii* in having much paler perianth segments and only 3–4 stamens.

5. **Juncus punctorius** L.f., Suppl. Pl.: 208 (1782). —Baker in Fl. Cap. **7**: 20 (1897); in F.T.A. **8**: 93 (1901). —Obermeyer in F.S.A. **4**(2): 79 (1985). —Clarke & Klaassen, Water Pl. Namibia: 82 (2001). —Cook, Aquat. Wetl. Pl. Sthn. Africa: 155 (2004). Type: South Africa, Cape of Good Hope, *Thunberg & Sonnerat* s.n. (LINN 449.15 holotype).

Juncus exaltatus Decne. var. *capensis* Kunth, Enum. Pl. **3**: 329 (1841). Type: South Africa, Cape of Good Hope, *Drège* s.n. (E isotype).

Juncus schimperi A. Rich., Tent. Fl. Abyss. **2**: 338 (1851) Type: Ethiopia, Chire Province, near Adoam (Adua), 1.xii.1837, *Schimper* 56 (P holotype, K, W).

Perennial herb, 40–110(220) cm tall. Rhizome creeping, woody; roots numerous and thin. Stems rigid. Leaves basal, single cauline leaf inserted in upper part of stem, cataphylls 2–3, sheaths auriculate; blade terete, septate. Inflorescence a panicle, subtending bract shorter than the inflorescence; capitula compact, semi-globose to globose, 10–80 per panicle, each with 8–30 flowers; floral bract solitary. Tepals ovate, acute to apiculate, light straw-brown, equal, 2–2.5 mm long. Stamens 3(6). Capsule ovoid, apiculate, shiny, light to dark red-brown, 3-angled, ± 0.5 mm long. Seeds c.0.45 mm long, ovoid, longitudinally striate, apiculate.

Zimbabwe. S: Masvingo Dist., 40 km E of Masvingo (Fort Victoria), x.1930, *Fries, Norlindh & Weimarck* 2148 (LD) (cited in Kirschner *et al.* 2002).

Also in Kenya, Lesotho, Namibia and South Africa. In permanently wet places, swamps, marshes and riverbanks; sea-level to 2200 m.

Conservation notes: Widespread species; not threatened.

No specimens seen from the Flora area. All measurements are taken from the literature.

6. **Juncus oxycarpus** Kunth, Enum. Pl. **3**: 336 (1841). —Baker in Fl. Cap. **7**: 20 (1897); in F.T.A. **8**: 93 (1901). —Carter in F.T.E.A., Juncaceae: 3 (1966). —Haines & Lye, Sedges Rushes E Africa: 34 (1983). —Obermeyer in F.S.A. **4**(2): 80 (1985). —Clarke & Klaassen, Water Pl. Namibia: 82 (2001). —Cook, Aquat. Wetl. Pl. Sthn. Africa: 154 (2004). Types: South Africa, Cape Province, Liesbeek R., *Bergius* s.n. (B† syntype); Paarl, Berg R., *Drege* a (K, P syntypes). FIGURE 13.4.15.

Juncus quartinianus A. Rich., Tent. Fl. Abyss. **2**: 339 (1851). Type: Ethiopia, Chiré, *Quartin-Dillon & Petit* s.n. (P holotype).

Juncus brevistilus Buchenau in Abh. Naturwiss. Vereine Bremen **4**: 433 (1875). Type: South Africa, Cape of Good Hope, locality & collector unknown.

Juncus gentilis N.E. Br. in Bull. Misc. Inform., Kew **1914**: 83 (1914). Type: South Africa, Gauteng, Modderfontein, 1896, *Conrath* 1173 (K holotype).

Juncus oxycarpus subsp. *microcephalus* Adamson in J. Linn. Soc., Bot. **50**: 13 (1935). Types: South Africa, Cape Province, Riversdale, *Muir* 3385 (BOL syntype); Grahamstown, *Dyer* 173 (BOL syntype).

Juncus suboxycarpus Adamson in J. Linn. Soc., Bot. **50**: 14 (1935). Type: South Africa, KwaZulu-Natal, Clairmont, *Schlechter* 3043 (BOL holotype, PRE).

Juncus oxycarpus subsp. *sparganioides* Weim. in Svensk Bot. Tidskr. **40**: 166 (1946). Type: Kenya, Mt Kenya, Liki, 11.ii.1922, *Fries & Fries* 1477 (UPS holotype, LD).

Erect perennial herb, sometimes decumbent, 10–80 cm tall. Rhizome short, creeping; roots thin, many, densely caespitose. Stems terete. Leaves basal, 2–3 cauline leaves per stem, cataphylls present, sheaths auriculate, green or grey-brown, sometimes with red tinge; blades terete, septate. Inflorescence a loosely branched terminal panicle, subtended by a bract; capitula 3–20, compact, semi-globose to globose, each with 4–30 flowers, each subtended by single triangular bract. Tepals lanceolate to attenuate, green to brown when immature, turning dark red-brown, margins membranous; inner tepals 3–3.6 mm long, outer tepals 2.8–3.3 mm long. Stamens 3(6). Capsule 2.5–3.5 mm long, oblong, with apiculate apex, 3-angled, shiny, light brown to dark red-brown, as long as perianth. Seeds 0.3–0.5 mm long, ovoid, apiculate, pale brown, reticulate, appendages absent.

Botswana. SE: Gaborone, Aedume Park, Gaborone Dam, 22.x.1977, *Hansen* 3246 (BM, C, GAB, K, PRE, SRGH, UPS, WAG). **Zambia**. N: Mpika Dist., dambo 97 km S of Mpika, 16.x.1967, *Simon & Williamson* 1045 (K, LISC, PRE, SRGH). W: Ndola, 5.ix.1967, *Mutimushi* 2040 (K, NDO). C: Mkushi Dist., Fuvela, 10.i.1958, *Robinson* 2719 (K, SRGH). **Zimbabwe**. W: Matobo Dist., Besna Kobila Farm, x.1956, *Miller* 3713 (K, PRE, SRGH). C: Harare, Kaola Estate, 11 km S of Harare (Salisbury), 1464 m, 14.x.1955, *Drummond* 4898 (K, LISC, PRE, SRGH). E: Nyanga Dist., Nyanga Nat. Park, Pungwe R. source, 26.i.1951, *Chase* 3641 (LISC, PRE, SRGH). S: Chivi Dist., near Madzivire Dip, 6.4 km N of Lundi R. bridge, 3.v.1962, *Drummond* 7883 (K, PRE, SRGH). **Malawi**. N: Rumphi Dist., Nyika Plateau, Lake Kaulime, 4.i.1959, *Robinson* 3030 (K, PRE, SRGH). C: Dedza Dist., 6.4 km from Dedza, 2.ii.1959, *Robson* 1426 (K, LISC). S: Ntcheu Dist., Kirk Range, Linsangwe R., 4.xi.1950, *Jackson* 278 (K). **Mozambique**. MS: Sussundenga Dist., Rotanda, Tandara, 1700 m, 19.xi.1965, *Torre & Correia* 13167 (BR, COI, LISC, LMA, LMU).

Also known from Angola, Kenya, Lesotho, Tanzania and South Africa. In swamps, marshes and along edges of streams, often forming large stands; sea-level to 3150 m.

Conservation notes: Widespread within the Flora region; not threatened.

Occasionally flowers develop into vegetative buds.

7. **Juncus exsertus** Buchenau in Abh. Naturwiss. Vereine Bremen **4**: 435 (1875). — Baker in Fl. Cap. **7**: 21 (1897). —Obermeyer in F.S.A. **4**(2): 80 (1985). —Adamson in J. Linn. Soc., Bot. **50**: 16 (1935). —Cook, Aquat. Wetl. Pl. Sthn. Africa: 153 (2004). Types: South Africa, Cape Province, Zwartkops R., *Zeyher* 103 (B† syntype, BOL); Worcester, Waterfall, *Ecklon & Zeyher* s.n. (B† syntype, PRE); Graaf-Reinet, Sundays R., *Bolus* 188 (BOL syntype).

Juncus rostratus Buchenau in Abh. Naturwiss. Vereine Bremen **4**: 437 (1875). Types: South Africa, Cape Province, Zwartkops R., xii.1829, *Ecklon & Zeyher* s.n. (S syntype); Bashee River, *Drège* 4465 (K, G syntypes).

Juncus exsertus subsp. *lesuticus* B.L. Burtt in Notes Roy. Bot. Gard. Edinb. **45**: 191 (1989). Type: South Africa, Cape, Barkley East, Witteberg, Ben McDhui, 4.ii.1983, *Hilliard & Burtt* 16435 (E holotype, NU, UPS).

Erect perennial herb, sometimes decumbent, 30–50 cm tall. Rhizome short, woody, creeping; roots thin, caespitose. Stems rigid. Basal leaves present or absent, 2–5 cauline leaves, cataphylls present, straw to brown in colour, sheaths auriculate; blades terete, septate. Inflorescence a lax panicle, appearing terminal due to short subtending bract, capitula 5–50, each with 3–15 flowers; floral bract solitary. Tepals unequal, inner 2.9–3.5 mm long, broadly lanceolate, acute to mucronate, margins broad, membranous; outer tepals 2.4–3 mm long, lanceolate, with darkened central band. Stamens 3–6. Capsule unilocular, 3.5–3.8 mm long, exceeding tepals, oblong, ovoid, slender, 3-angled, shiny, yellow to dark brown, apex narrow or beak-like. Seeds 0.4–0.6 mm long, ellipsoid, apiculate, light yellow brown, appendages absent, reticulate.

Zambia. C: Lusaka Dist., Munali, Little Dambo, 8 km E of Lusaka, 15.vi.1955, *Robinson* 1301 (K). **Zimbabwe**. N: Lomagundi Dist., Chapingabadza Farm, 20.xi.1969, *Jacobsen* 4061 (PRE). C: Harare (Salisbury), Thorn Park, 15.xii.1956, *Paterson* 16

(PRE, SRGH). E: Mutare Dist., tributary of Tsungwesi R. between Pounsley turnoff and Tungwesi bridge, 29.xi.1955, *Drummond* 5057 (K, LISC, SRGH). S: Umzingwane Dist., 10 km from Esigodini (Essexvale) on Bulawayo road, 4.i.1974, *Mavi* 1539 (K, SRGH).

Also in Lesotho, Namibia, South Africa and Swaziland. Along swamps, stream banks or in shallow water, sometimes forming large stands; 200–2600 m.

Conservation notes: Widespread species; not threatened.

Flowers may develop into vegetative buds forming new plants. *Juncus exsertus* is similar to *J. oxycarpus* but the capsule does not exceed the perianth.

8. **Juncus effusus** L., Sp. Pl.: 326 (1753). —Carter in F.T.E.A., Juncaceae: 2 (1966). —Haines & Lye, Sedges Rushes E Afr.: 34 (1983). —Obermeyer in F.S.A. 4(2): 77 (1985). —Cook, Aquat. Wetl. Pl. Sthn. Africa: 153 (2004). Type: Europe, "Habitat in Europae uliginosis" (LINN 449.3 lectotype).

Perennial herb, 50–150 cm tall. Rhizome branched, horizontal; roots thick, many. Stems terete, green-yellow, finely ridged. Leaves basal, typically leafless, cataphylls present, red-brown; blade cylindrical, ending in a sharp point, not septate. Inflorescence a pseudo-lateral panicle, subtending bract forming a continuation of the stem; panicle of 50–120 stalked or sub-sessile flowers. Tepals linear to acute, green to light brown, margin membranous; outer tepals 1.5–3 mm long, longer than inner. Stamens 3(6). Capsule 1.6–2.5 mm long, ovoid to ellipsoid, yellow to dark brown, equal to or slightly exceeding perianth. Seeds 0.2–0.3 mm long, ovoid, yellow-brown, reticulate, with no appendages.

Cosmopolitan, although more common in the northern hemisphere, widespread in Africa.

Subsp. **laxus** (Robyns & Tournay) Snogerup in Preslia **74**: 259 (2002).

Juncus oehleri Graebner in Bot. Jahrb. Syst. **48**: 506 (1913). Type: Tanzania, Masai/Mbulu Dist., Lake Ossirwa, 20.ii.1907, *Oehler* 499 (B holotype).
Juncus laxus Robyns & Tournay in Bull. Jard. Bot. État **25**: 252 (1955). Type: Congo, Kivu, Kundhuru-Ya-Tshuve, 15.ix.1934, *de Witte* 1976 (BR holotype).

Inflorescence lax, lower branches often pendulous; bract not constricted below inflorescence. Anthers, 0.4–0.5 mm long.

Zimbabwe. E: Nyanga Dist., Nyanga Nat. Park, Mare Dam., 6.i.1972, *Gibbs Russell* 1199 (K, SRGH).

Also in Burundi, Kenya, Rwanda, South Africa, Uganda and Congo. Near streams and rivers in upland rainforest, evergreen bushland and grassland, mainly in montane areas; 1350–3100 m.

Conservation notes: Widespread taxon; not threatened.

Five subspecies are recognised, with only one occurring in the Flora region.

MUSACEAE

by J.M. Lock and M.A. Diniz

Giant perennial herbs from a branched or unbranched corm-like rootstock. Leaves arising from apex of corm, spirally arranged, very large; leaf-sheaths elongated, densely imbricate and forming a cylindrical pseudostem; lamina oblong, with a strongly channelled midrib and very many pinnately-arranged, parallel lateral veins. Inflorescence terminal on the corm, growing up through the centre of the pseudostem and thus appearing to arise from its apex. Flowers unisexual, those on the proximal parts of the inflorescence female, on the distal parts male, borne in condensed groups subtended by spathaceous bracts. Calyx spathaceous, splitting down one side, with ± 3 teeth at the apex. Corolla lobes 3, two of them joined to the calyx tube, the third separate and directed downwards. Stamens 6, but one usually rudimentary; filaments terete, thin; anthers dithecous with parallel thecae. Ovary inferior, trilocular, placentation axile; ovules many. Fruit a large elongated, fleshy trilocular berry containing, in wild species, numerous very hard subspherical seeds with a straight embryo and copious endosperm.

The family contains two genera, *Musa* and *Ensete*. Only the latter definitely occurs wild in Africa (although *Musa acuminata* may be genuinely wild in Pemba Island, Tanzania), but cultivars of *Musa* (bananas and plantains) are very widely grown and in some places provide a staple diet. These cultivars and their origin are discussed further below.

Plants unbranched so that vegetative shoots are always solitary; each plant dying after flowering (monocarpic); wild plants with fruits containing large hard black seeds · **1. Ensete**
Plants branched below ground, producing a clump of vegetative shoots; each shoot dying after flowering; cultivated plants with seedless fruits · · · · · · · · · · **2. Musa**

1. **ENSETE** Horan.

Ensete Horan., Prodr. Monogr. Scitam.: 40 (1862). —Simmonds in Kew Bull. **14**: 198–212 (1960).

Perennial giant herbs from an unbranched corm-like base. Whole plant dying after flowering. Leaves very large; sheaths imbricate, forming a pseudostem.

About 7 species in Africa, Madagascar, India and SE Asia. Baker & Simmonds (Kew Bull. **8**: 405–416, 1953) recognize only 3 species in Africa. In view of the difficulty of obtaining measurements from dried material, we have drawn heavily on the general description given by Baker & Simmonds (1953), prepared largely from fresh material.

1. Plant generally 5–9 m high, in forest; leaves aggregated towards apex of a perennial aerial pseudostem; seeds 10–23 mm at largest diameter · · · · · · · · **1.** *ventricosum*
– Plant generally smaller, 3 m high or less, in savanna; leaves in a ± sessile rosette, dying back to a basal corm in the dry season; leaves spaced out along the flowering stem; seeds up to 9 mm at largest diameter · · · · · · · · · · · · · · · · · · 2
2. Lateral nerves of leaf forming an angle of 50–70° to midrib; plant 1.5–3 m high when flowering, at least sometimes dying back to a basal corm in dry season; seeds 7–9 mm in diameter; upland grasslands and woodlands, often with *Pteridium* · **2.** *gilletii*
– Lateral nerves of leaf forming an angle of 45° to midrib; plant no more than 1.5 m high when flowering, dying down to a corm-like structure in dry season; seeds 6–7 mm in diameter; Zambezian woodlands, often around termitaria · **3.** *homblei*

1. **Ensete ventricosum** (Welw.) Cheesman in Kew Bull. **2**: 101 (1948). —Lock in F.T.E.A., Musaceae: 3 (1993). —White *et al.*, Evergr. For. Fl. Malawi: 103 (2001). —M. Coates Palgrave, Trees Sthn. Africa, ed. 3: 117 (2002). Type: Angola, Pungo Andongo, iii–v.1857, *Welwitsch* 6447 (LISU holotype, BM, K). FIGURE 13.4.**16**.

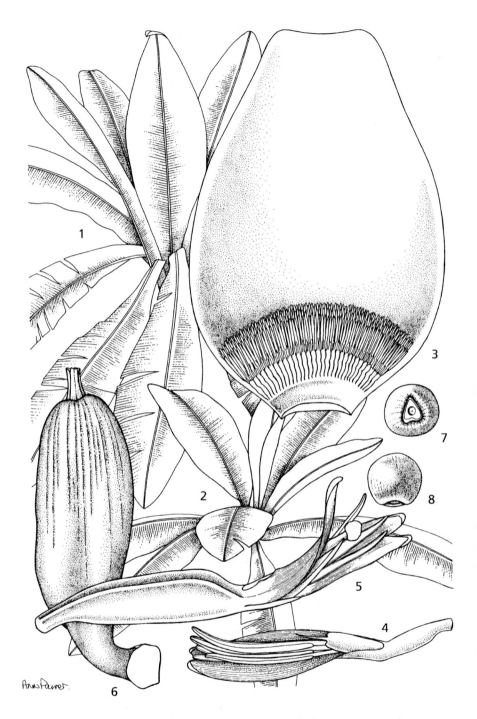

Fig. 13.4.**16**. ENSETE VENTRICOSUM. 1 & 2, habit (from photo); 3, bract (× ¹/₅); 4, male flower (× 1); 5, female flower (× 1); 6, fruit (× ²/₃), 7, 8 seeds (× 1). 1–3, 6–8 from *Baker & Simmonds* 114, 4–5 from *Baker & Simmonds* 113. Drawn by Ann Farrer. From Flora of Tropical East Africa.

Musa ensete J.F. Gmel., Syst. Nat., ed.13 **2**: 567 (1791). —Baker in F.T.A. **7**: 329 (1898). —Brenan, Checklist For. Trees & Shrubs Tanganyika: 364 (1949). Type: Ethiopia, description and plates in Bruce, Travels Nile, Egypt, Arabia, Abyssinia & Nubia **5**: 36–41 (1790).

Musa ventricosa Welw. in Ann. Cons. Ultram., parte não official, sér.1: 587, no.45 (1859). —Ridley in J. Bot. **25**: 134 (1887). —Baker in F.T.A. **7**: 330 (1898).

Ensete edule Horan., Prodr. Monogr. Scitam.: 40 (1862). Type as for *M. ensete*.

Musa buchananii Baker in Ann. Bot. **7**: 207 (1893); in F.T.A. **7**: 329 (1898). Type: Malawi, Shire Highlands, 610 m, vii.1885, *Buchanan* 470 (K holotype).

Ensete buchananii (Baker) Cheesman in Kew Bull. **2**: 102 (1948).

Musa davyae Stapf in Bull. Misc. Inform., Kew **1913**: 103 (1913). Types: South Africa, Soutspansberg, *Burtt Davy* (drawings at K) and Mozambique, Amatongas Forest, 27.xi.1907, *W.H. Johnson* 54 (K).

Giant herb arising from a short upright rhizome. Pseudostem formed of overlapping leaf-bases, 1.5–7(9) m tall. Leaf-blades erect or spreading, forming a large rosette, oblong-lanceolate, to 5 × 1.5 m, glaucous or not, midrib red or green. Inflorescence appearing from centre of rosette, pendulous when mature. Bracts of male part of inflorescence persistent or partially deciduous, up to 55 × 18 cm, each subtending ± 30–40 flowers. Calyx of male flowers 3-lobed, lobes variable in length, 3.5–5.5 cm long, white with orange-yellow tips; corolla serrate-apiculate, 1–1.5 × 1–1.7 cm, apiculum 0.3–1.3 cm or occasionally absent; stamens 5, 3–5 cm long, anthers violet to purple, filaments white; staminode present or not, needle-like, 0.1–1 cm long; style needle-like, 1–2 cm long. Bracts of female or hermaphrodite part of inflorescence persistent, partially covering fruits. Calyx spathaceous, 3-lobed, sometimes with 1–2 smaller extra narrow pointed lobes attached to it internally; petals 1–3, variable in shape with 2 wings and an apiculum up to 1.5 cm long; stamens 0–5, 3.5 cm long, coloured as in male flowers; staminodes variable according to number of stamens present; style 2.5–4 cm long, terete, with a large capitate stigma. Fruits 5–20 in axil of each bract, long-obovoid, (7)8–15 × 3–4.5 cm, orange at maturity. Seeds irregularly subspherical, 1.2–2.3 × 1.2–1.8 × 0.9–1.6 cm, striate to smooth, hard, black, embedded in orange pulp.

Zambia. Recorded from Northern and North-western Provinces in Phiri, Checklist Zambia Vasc. Pl. (2005), but no specimens seen. **Zimbabwe.** N: Bindura Dist., Chinamora communal land (Chindamora Reserve), 1200 m, fl. 15.iv.1921, *Eyles* 3431 (SRGH). C: Goromonzi Dist., Umwinzi Valley, E of boundary of Greystone Park, planted, 21.vi.1965, *West* 6597 (K, LISC, SRGH). E: Chimanimani Dist., Bridal Veil Falls, 5 km NW of Chimanimani (Melsetter), 1200–1500 m, fl.& fr. 30.xii.1950, *Crook* 342 (BM, K, PRE). **Malawi**. N: Rumphi Dist., Nyika Plateau, c.35 km N of Ml, 6.xi.1977, *Pawek* 13209 (K). C: Dedza Dist., Dedza Mt, fl. 10.iv.1980, *Blackmore, Banda & Patel* 1225 (BM, K). S: Thyolo Dist., Thyolo (Cholo) Mt, 1200 m, 25.ix.1946, *Brass* 17795 (K). **Mozambique**. N: Malema Dist., Serra Murripa, 40 km from Malema (Entre-Rios) to Ribaué, fl. 15.xii.1967, *Torre & Correia* 16535 (BR, COI, LISC, LISU, LMA, LMU, MAL, PRE, SRGH, WAG). MS: Sussundenga Dist., Mavita, R. Moçambize, 1300 m, fl. 24.x.1944, *Mendonça* 2597 (BR, LISC, LMU, PRE).

Also in Uganda, Kenya, Tanzania, Ethiopia and Cameroon south to Angola and South Africa (former Transvaal). Disturbed places in upland forest, often in ravines and on steep slopes, or in swamps and on river banks, sometimes also in drier lowland forests; 800–2250 m.

Conservation notes: Widespread in tropical African montane and sub-montane forests, often apparently benefiting from disturbance; not threatened.

Undoubtedly under-collected because of its size and because of confusion with cultivated bananas. It is sometimes planted as an ornamental (Biegel, Checkl. Orn. Pl. Rhod. Gard., 1977).

E. ventricosum forms the basis of an agricultural system in S Ethiopia. The corms are harvested just before flowering, and the stored starch is extracted. The system is described by Smeds in Acta Geogr., Helsinki **13**: 1–40 (1955) and by Westphal in Belmontia N.S. **14**: 123–163 (1975).

2. **Ensete gilletii** (De Wild.) Cheesman in Kew Bull. **2**: 103 (1947). —Baker & Simmonds in Kew Bull. **8**: 407 (1953). —Hepper in F.W.T.A. ed.2, **3**: 69, fig.341. Type: Congo, Bas-Congo, Kisantu, 1900, *Gillet* 700 (BR).

    Musa gilletii De Wild. in Rev. Cult. Colon. **8**: 102 (1901); in Not. Pl. Ut. Intér. Fl. Congo **1**: 73, t.5, 6 (1903).

    Musa livingstoniana Kirk in J. Linn. Soc, Bot. **9**: 128 (1867). —Baker in F.T.A. **7**: 330 (1898) in part as regards type. Type: Seeds collected in Malawi, Zomba, Maganja Hills, 880 m, xi.1903, by J. Mahon, sent to Baker at K.

    Ensete livingstoniana (Kirk) Cheesman in Kew Bull. **2**: 101 (1947) in part as regards type.

Perennial monocarpic herb 1.5–3 m high. Young stems dying back to a hard round corm c.30 cm in diameter at soil surface during the dry season. At flowering, leaf-blades spreading, spaced on pseudostem, the lowest up to 150 cm long, elliptic to narrowly elliptic, apex acute to acuminate, base cuneate. Midrib prominent below; lateral nerves numerous, forming an angle of 50–70° with midrib. Inflorescence terminal on leafy shoot. Inflorescence bracts 4.5–9 × 17–25 cm, very broadly ovate, male ones mauve. Male flowers: calyx spathaceous, splitting down one side, remaining lobe ligulate, 3-lobed at apex; dorsal petal 3-lobed, with broad rounded lateral lobes and a central apiculus; stamens 6, 12 mm long. Female flower as male but ovary much larger and stamens rudimentary. Fruit c.5 × 2 cm, narrowly ellipsoid, obtuse, splitting at maturity to reveal seeds embedded in orange pulp. Seeds 7–9 mm diameter, black, hard.

Malawi. S: "Tohinsunze, Shirwa", iv.1859, *Kirk* s.n. (K).

Occurs scattered from Sierra Leone to Malawi, usually in drier habitats than *E. ventricosum*, and usually at higher altitudes than *E. homblei*.

Conservation notes: A poorly known taxon, almost certainly overlooked and under-collected. Best treated as Data Deficient, but further research may show that a rating of Vulnerable is more appropriate.

Baker (1898) mentions that the species (as *Musa livingstoniana*) occurs in Gorongosa, Mozambique. However, we have seen no material from that country.

The description of *Musa livingstoniana* was published by Kirk in 1867, who states "Hab. the mountains of Equatorial Africa. Gorongozo, south latitude 19°; Manganja, south latitude 14°; Maravi country, south latitude 12°, and the Niger region?" without any reference to collected material. Cheesman (1947) also does not cite any type material. Baker & Simmonds (1953, p.408) refer to the type of *E. livingstonianum* as "J. Mahon, Zomba, Nyasaland, in Herb. Kew", and reject the species because they consider *M. livingstoniana* a *nomem confusum*. In fact, the species description by Kirk refers to *E. ventricosum* but the seeds refer to *E. gilletii*.

The relationship between this species and *E. homblei* deserves further study in the field. It possible that they are forms of a single variable species; the differences appear to be mainly quantitative.

3. **Ensete homblei** (De Wild.) Cheesman in Kew Bull. **2**: 103 (1947). —Milne-Redhead in Hooker, Icon. Pl. **35**: t.3479 (1950). —Baker & Simmonds in Kew Bull. **8**: 408 (1953). Type: Congo, Katanga, around Lubumbashi (Elizabethville), on termitaria, v.1912, *Homblé* 671 (BR holotype). FIGURE 13.4.17.

    Musa homblei De Wild. in Ann. Inst. Bot.-Géol. Colon. Marseille, sér.2, **10**: 332 (1912).

Perennial monocarpic herb to 2.4 m high from a swollen base covered with brown scales. Pseudostems annual, formed by overlapping leaf-bases. Leaf-blades spreading, spaced on pseudostem, 17–40 × 7–11.5 cm, elliptic, apex acute to obtuse, sometimes apiculate, base cuneate, green above, glaucous beneath. Leaves of non-flowering individuals larger than those of flowering plants. Inflorescence terminal, ovoid, dense, compact, to 13 × 8 cm. Inflorescence bracts to 7 × 5 cm, ovate, concave, apex obtuse, base contracted, the lower 9–10 subtending female flowers, the middle 1–2 hermaphrodite flowers, and the upper ones, male flowers. Female flowers 3.5 cm long. Calyx forming a tube 18 × 9 mm, trilobed at apex and

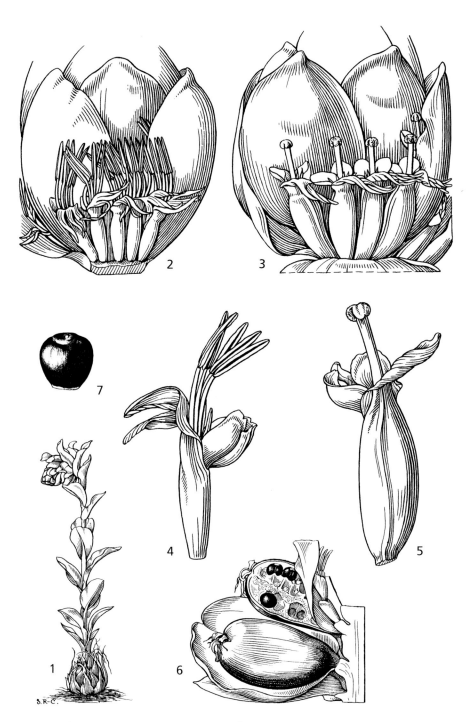

Fig. 13.4.**17**. ENSETE HOMBLEI. 1, habit ($\times$ $^1/_4$); 2, inflorescence, some bracts removed to show male flowers ($\times$ 1); 3, inflorescence, some bracts removed to show female flowers ($\times$ 1); 4, male flower ($\times$ 2); 5, female flower ($\times$ 2); 6, part of perianth, base of filaments and gynoecium ($\times$ $^2/_3$); 7, seed ($\times$ 2). Drawn by Stella Ross-Craig. Adapted from Hooker's Icones Plantarum, t.3479.

longitudinally split dorsally, reflexed and spirally twisted after anthesis; lobes c.8 mm long, obtuse. Dorsal corolla lobe 3-lobed, c.8 × 13 mm, the other two fused to calyx. Staminodes 5. Style 12 mm long × 1.5 mm diameter; stigma 6-lobed; ovary subcylindric, c.18 mm long. Male flowers similar to female, but lateral lobes of corolla free, c.9 mm long, very narrowly ovate; stamens 5, filaments c.10 mm long; anthers c.9 mm long; staminodes 1, minute; vestigial ovary present, c.10 mm long. Fruits narrowly pear-shaped, somewhat angled, to 4.5 × 2.3 cm, 20–30-seeded. Seeds subglobose, 6–7 mm in diameter, black, shiny, enclosed in white pulp; hilum round, c.4 mm in diameter.

Zambia. N: Mansa Dist., Mansa (Fort Rosebery), 7.xii.1962, *Baker* M13 (K, SRGH); Kawambwa, fl. 9.xi.1957, *Fanshawe* 3901 (K). W: Mwinilunga Dist., Kalenda Ridge, W of Matonchi Farm, 2.xi.1937, *Milne-Redhead* 3062 (K); Kabompo Dist., 6.5 km W of Kabompo Gorge, fl. 25.x.1966, *Leach & Williamson* 13476 (LISC, SRGH).

Also in S Congo (Katanga). On or around termite mounds in *Brachystegia* woodland; 1000–1200 m.

Conservation notes: The species range is fairly restricted, as is its termite mound habitat. The apparently monocarpic habit also makes it somewhat vulnerable to stochastic changes. However, without more knowledge of its distribution, ecology and abundance, it is best treated as Data Deficient.

A very small 'banana', flowering only once and then dying. It apparently grows for several years before flowering, dying down to a corm-like structure in the dry season. Milne-Redhead (1950) states that suckers are produced at the base of the plant, as do field-notes on *Baker* M13, and *Schmitz* 4766 (K) from S Congo may show them, albeit detached. Baker & Simmonds (1953) found that apparent suckering in *E. ventricosum* was due to groups of seeds germinating together. Further careful observations on *E. homblei* are needed.

2. **MUSA** L.

Musa L., Sp. Pl.: 1043 (1753); Gen. Pl., ed.5: 466 (1754). —Stover & Simmonds, Bananas, ed.3 (1987).

Perennial giant herbs from a branching corm-like rhizome; each shoot dying after flowering. Leaves very large; sheaths imbricate, forming a pseudostem up to 4 m high; lamina oblong.

With the possible exception of *M. acuminata* from Pemba Island, Tanzania, there are no wild species of *Musa* in Africa. The common hybrid cultivars described below are, however, widespread in the moister parts of the Flora area. They are widely eaten fresh or dried, cooked or not, and also used to prepare alcoholic drinks.

Musa cultivars

Simmonds has shown that the cultivated bananas are of SE Asian origin. They are sterile diploids, triploids or tetraploids derived either from *M. acuminata* Colla alone, or from hybrids with *M. balbisiana* Colla. Varieties have originated by selection from these hybrids, and subsequently by somatic mutation and selection.

Simmonds & Shepherd (J. Linn. Soc., Bot. **55**: 302–312, 1955) showed that both Linnaean names (*M. sapientum* and *M. paradisiaca*) are based on hybrid cultivars, and that neither can be correctly used for the cultivated bananas. They proposed a system of cultivar classification within the *Eumusa* series based on ploidy and genome composition. Thus a diploid plant containing genetic material from only *M. acuminata* is referred to as *Musa* (AA Group) '[cultivar name]', and a triploid plant containing material from both species is referred to as *Musa* (AAB Group) '[cultivar name]', and a tetraploid, also with genetic material from both species, as *Musa* (ABBB) '[cultivar name]'. This scheme falls within the confines of the International Code of Nomenclature for Cultivated Plants and combines simplicity, precision and informativeness. By using a number of morphological and colour characters, Simmonds & Shepherd devised a scoring system whereby the genome composition of a cultivar could be determined with some certainty; they also describe how, with experience, the degree of ploidy of a cultivar may be estimated in the field. This Flora is not concerned with the identification of cultivars; those with a special interest are referred to the original paper or to the book on bananas by Stover & Simmonds (1987).

STRELITZIACEAE

by M.A. Diniz & J.M. Lock

Stemless or caulescent herbs, usually perennial, often with a horizontal rootstock, sometimes tall and tree-like. Leaves large, distichously arranged at apex of stem or basal, petiolate; leaf-sheaths forming a cylindrical stem. Inflorescence terminal at stem apex, long-pedunculate with 1 or various reddish-brown or claret-coloured boat-shaped, coriaceous, spathaceous bracts including the flowers. Flowers bisexual, zygomorphic, showy; calyx with 3 subequal white, cream or yellow-orange sepals; corolla with 3 petals, the 2 inferior equal and united to form a sagittate blade enclosing stamens and style, blue or white, the third much smaller. Stamens 5, filaments terete, thin, rigid; anthers linear, dithecous. Ovary inferior, tricarpellary, trilocular, with axial placentation; style filiform, stigma 3-lobed. Fruit a 3-lobed capsule, 3-valved, loculicidally dehiscent; seeds few, globose, with or without albumen, arillate.

A family with 3 genera and about 7 species occurring in South America (*Phenakospermum*), southern Africa (*Strelitzia*) and Madagascar (*Ravenala*). *Ravenala* and *Strelitzia* are frequently cultivated as ornamentals in parks and gardens.

STRELITZIA Aiton

Strelitzia Aiton, Hort. Kew. **1**: 285 (1789).

Perennial stemless plants or with tree-like stems arising from a rhizome with many fleshy adventitious roots. Leaves distichous, usually sheathing at base, long-petiolate, lamina large to very large, oblong to oblong-lanceolate, dark green and leathery, channelled on lower surface. Peduncle long, cylindrical, ending in a short condensed inflorescence enclosed in a boat-shaped coriaceous, green or reddish-brown spathaceous bract. Flowers bisexual, irregular. Calyx petaloid with 3 sepals, lower one boat-shaped towards base, attenuate above, keeled, 2 superior ones attenuate. Petals 3, white, blue or orange-blue, unequal; 2 lower petals equal, united into a sagitte segment with central channel enclosing stamens and style; upper petal much smaller. Stamens 5; filaments terete, thin, rigid; anthers linear, dithecous, with longitudinal slits. Ovary inferior with 3 carpels, trilocular, placentation axial, ovules many in each locule, rarely solitary; style filiform with 3 linear stigmatic branches. Fruit a hard 3-lobed capsule, 3-valved, loculicidally dehiscent; seeds few, globose, with woolly aril.

About 5 species in Africa, predominantly in southern Africa; 2 species in the Flora Zambesiaca area.

We have had difficulty in obtaining measurements from dried material. Furthermore, herbarium specimens of parts of leaves are more common than those with flowers; such very incomplete material is extremely difficult to identify correctly. Vegetatively the species are very similar; when without flowers they are easily confused.

Fig. 13.4.**18**. STRELITZIA CAUDATA. 1, habit; 2, part of leaf showing venation (× ²/₃); 3, detail of venation on upper leaf surface (× ¹/₃); 4, spathe with two emerging flowers (× ¹/₂); 5, flower and three sepals (× ²/₃); 6, two joined petals with stamens and style within (× ²/₃); 7, anther (× 1); 8, style (× ²/₃). 2–5, 8 from *Heenan* 31932, 5–6 from *Scheepers* 428, 8 from *Crook* M99. Drawn by Juliet Williamson.

Inflorescence simple; lower petals united, with short obtuse lobes at base · · · · ·
· **1.** *caudata*
– Inflorescence compound; lower petals united, with long sagittate lobes at base ·
· **2.** *nicolai*

1. **Strelitzia caudata** Dyer in Fl. Pl. Afr. **25**: t.997 (1946). —Palmer & Pitman, Trees
   Sthn. Africa **1**: 409 (1972). —M. Coates-Palgrave, Trees Sthn. Africa, ed.3: 118
   (2002). Type: South Africa, Limpopo Prov., Zoutpansberg Dist., Farm Geluk,
   c.22 km E of Louis Trichardt, *Verschuur* s.n. in PRE 21,654 & 27,162 (PRE
   syntypes). FIGURE 13.4.**18**.

Tree-like perennial, unbranched, usually to 2 m high but may reach 10 m, 10–15 cm in
diameter, producing suckers from the base. Leaves distichously arranged, forming a tuft at
apex; petiole 1.5–1.8 m long; lamina ovate to oblong, up to 2 m long and 45–80 cm wide,
cordate at base. Inflorescence terminal; peduncle 30 cm long or more, 1.5–2 cm wide,
spreading at apex; bracteate, pink. Bracts up to 45 cm long and 10 cm wide, surrounding the
peduncle, lanceolate, powdery-glaucous. Spathe solitary, boat-shaped, at right angles to
peduncle, coriaceous, 25–30 × 6–6.5 cm, c.3.5 cm thick at broadest, claret-coloured, glaucous,
tapering to a slender tip, exuding mucilage from near the base, subtending many flowers,
exserting successively from spathe; each flower surrounded by a claret-coloured bract. Sepals
3, white, sometimes mauve-tinged at base, 2 superior widely diverging from lower, linear-
lanceolate, 12.5–19 × 2.5–3.5 cm at widest towards base, slightly concave above, strongly
keeled below; inferior about equal in length to others, boat-shaped, strongly keeled,
sometimes keel projecting around middle into a slender lobe or segment 1.5–2.5 mm. Petals
3, light mauve throughout, sometimes only towards base, 2 lower ones united, 10–16 cm long,
basal parts forming a boat-shaped depression 2–3 cm wide, with undulate margins and
crisped; upper parts forming a sagittate lamina, 8–10 × 2 cm, with obtuse basal lobes 5–6 ×
7–8 mm; upper petal 2.5–3 × 1–1.25 cm, ovate, cuspidate, slightly keeled down inner face,
concave. Filaments terete, 3–4 cm long; anthers 5–9 cm long. Style 10–14 cm long, tapering
into stigmatic branches, 3.5–5 cm long, viscid. Ovary 3–5 cm long, irregularly triangular in
cross-section. Capsule 4–6 × 3.5 cm, triangular across sides, hard. Seeds with a tuft of yellow
hairs (red at maturity).

Zimbabwe. E: Chimanimani Mts, fl.& fr. 3.vi.1948, *Munch* 127 (K, SRGH).
Mozambique. MS: Gorongosa, Mt Nhandore summit, 1840 m, fl. 19.x.1965, *Torre &*
Pereira 12434 (BR, COI, K, LISC, LMA, LMU, SRGH).
   Also in Swaziland, South Africa (former Transvaal). In rocky places, ravines and in
savanna grasslands usually in moist rich soils; 1200–2100 m.
   Conservation notes: Restricted montane distribution in the Flora area; probably
Vulnerable.
   Cultivated in gardens and parks as an ornamental. Called 'wild banana' in
Chimanimani area of Zimbabwe. In Gorongosa, Mozambique, the stem pith is
considered edible.

2. **Strelitzia nicolai** Regel & Körn. in Gartenflora **7**: 265, t.235 (1858). —Wright in Fl.
   Cap. **5**(3): 318 (1913). —Dyer in Fl. Pl. Afr. **25**: t.996 (1946). —M. Coates-
   Palgrave, Trees Sthn. Africa, ed.3: 119, t.13 (2002). Type: Plant grown in
   Imperial Gardens of Emperor Nicolas at St. Petersburg, Russia; of unknown
   origin but probably South Africa between Durban and East London.
      Strelitzia augusta sensu Wright in F.C. **5**(3): 318 (1913) in part for Natal specimens.

Tree-like perennial herb up to 10 m tall, producing suckers from base and clump-forming.
Leaves arranged distichously in tufts at apex; petiole up to 2 m long; lamina oblong or ovate-
oblong, to 1.5 × 0.6 m, rounded or rarely cordate at base, shiny green. Inflorescence at stem

apex, many-flowered with a horizontal peduncle 30–45 cm long or more, bracteate; bracts lanceolate, surrounding peduncle, the uppermost one exceeding it, long-acuminate, glaucous. Spathe boat-shaped, spreading-ascending, 40–45 cm long, 8 cm high towards base, 3.5–4 cm thick, acuminate, coriaceous, claret-coloured, glaucous, sheathing inflorescence, followed inside by a second and third spathe, sometimes up to five spathes sheathing secondary inflorescences produced from the principal axis; inner spathes and flowers smaller than outer ones. Sepals 3, white, sometimes bluish towards base, the 2 superior lanceolate, 12–18 × 3 cm, keeled below, concave or channelled above; the lowest one boat-shaped, keeled towards base, slightly shorter than others. Petals 3, light blue to almost white, 2 lower ones 10–12 cm long, united to form a boat-shaped structure towards base, 3.5–4.5 × 1.5 cm, margins curved and united above middle, forming a sagittate lamina 8–10 × 2.5 cm, basal lobes 2–3 × 0.6–0.8 cm; upper petal cuspidate, c.15 × 7 mm wide with cusp c.5 mm long, acute, slightly keeled down inner face. Filaments c.4 cm long, slightly coiled; anthers to 8 cm long, slender. Style 9–12 cm long, stigmatic branches 3–6 cm long. Ovary c.5 cm long. Capsule triangular in cross-section, c.6 cm long and 2.5 cm wide across angles. Seeds oblong, 1 × 0.5 cm, with tuft of orange-coloured hairs.

Zimbabwe. E: Chipinge Dist., Tarka Forest Reserve, 38 km S of Chimanimani (Melsetter), banks of Chambuka (Chambuca) R., 1140 m, fl. ix.1951, *A.O. & L. Crook* s.n. (K, SRGH). **Mozambique**. M: Matutuíne Dist., between Zitundo and Ponta do Ouro, riverine forest of rio Cele, fl. 3.xii.1968, *Balsinhas* 1421 (LISC, LMA).

Also in South Africa. In riverine forest and banks of rivers, forming small colonies; up to 1140 m.

Conservation notes: Restricted distribution in Flora area; probably Vulnerable.

Dogs like to eat the plants, but the seeds are poisonous and when eaten can cause abdominal pains and vomiting.

COSTACEAE

by M. Lock & M.A. Diniz

Perennial, terrestrial or (not in the Flora area) epiphytic rhizomatous herbs. Sterile and fertile shoots separate or not. Vegetative parts without oil cells, not aromatic. Leaves either solitary (not in Flora area) or in a basal rosette, or arranged in an open spiral on the elongated leafy stem; sheaths closed, encircling stem; ligule encircling stem, sometimes termed an ochrea. Lamina usually elliptic, narrowing to a short false petiole, with a strong median nerve and parallel laterals diverging from midrib at a c.45° angle, themselves joined by parallel tertiary veins. All parts glabrous, or pubescent with bicellular hairs. Inflorescences on specialised bracteate shoots arising from rhizome, separate from the leafy shoots, or terminating the leafy shoots (both conditions may occur in a species), racemose, condensed into cone-like spikes formed of 1–2-flowered units, each subtended by a bract. Flowers large, strongly zygomorphic, each subtended by a bracteole (if 2 flowers present, one may be rudimentary). Calyx tubular, with 3 apical teeth, sometimes split to base on one side. Petals 3, all similar, or the posterior ± larger than laterals; basal third of petals and base of androecium fused to form a tube. Androecium of one broad petaloid staminode (labellum) and a single free stamen; stamen petaloid, usually elliptic, entire; anthers subterminal, the pollen-containing parts (anther-thecae) dehiscing longitudinally and completely. Labellum very broad and petaloid, variously coloured, curved into a funnel-like structure. Style filiform, passing between anther-thecae, ending in funnel-shaped stigma with ciliate margin and a dorsal appendage. Nectaries embedded in apex of ovary. Fruits bacciform or capsular, probably indehiscent (dehiscent in South American species), in African species usually remaining covered by bracts, crowned by a persistent calyx. Seeds hard, blackish, arillate.

Of the four genera in the family, only *Costus* occurs in Africa. The genus is, however, most diverse in South America (see Maas, Flora Neotropica 18, 1977). *Tapeinochilus* occurs in South-East Asia, and *Monocostus* and *Dimerocostus* in South America.

The family was formerly considered as part of Zingiberaceae, and recent analyses retain the two as sister groups. Costaceae differs from Zingiberaceae, among other features, in the phyllotaxy (spiral, not distichous), in the lack of oil cells, in the form of the stamen, and in the labellum which is formed from 5 (not 2) staminodes.

Two species, *Costus spectabilis* and *C. macranthus*, are very different in appearance from others in the genus, being rosette-forming savanna herbs that arise from deep underground rhizomes. The inflorescence of yellow flowers arises from the middle of a rosette of (usually) 4 leaves. These species have been distinguished as the genus *Cadalvena* Fenzl, but are here considered as part of *Costus*.

COSTUS L.

Costus L., Sp. Pl.: 2 (1753). —Schumann in Engler, Pflanzenr. IV. **46**: 378 (1904).
Cadalvena Fenzl in Sitzungsber. Kaiserl. Akad. Wiss., Math.-Naturwiss. Cl., Abt.1 **51**: 139 (1865).

Features as for the family, but with a consistently trilocular ovary; in the other genera the ovary is usually bilocular.

There are c.150 species of *Costus* in the New and Old World Tropics; with about 25 species in Africa, much in need of detailed revision.

The majority of the species occur in forests, often along streams or at forest margins; some are epiphytic. Two occur in open grasslands and woodlands, often on shallow soils over rock, or around termitaria. These two species are widespread (see distribution map in Lock, Kew Bull. **39**: 841, 1984); many of the other species, particularly epiphytic forest ones, are poorly collected, local in distribution, and much of need of revision, currently being undertaken by Maas and Maas. The flowers are very delicate and the study of herbarium specimens, unless very well prepared, is difficult.

1. Plant consisting of rosette of 4 leaves, appressed to the ground, or leafless when flowering; flowers yellow, arising from middle of rosette · · · · · · · · · · · · · · · 2
– Plant with elongated stem on which leaves are arranged in an open spiral; flowers in cone-like inflorescence at end of leafy shoot, white or pink with yellow markings · 3
2. Anther-thecae 4–7 mm long; petals 2–4.5 cm long; total length of dried flower less than 10 cm · **1.** *spectabilis*
– Anther-thecae 9–13 mm long; petals 6–8 mm long; total length of dried flower more than 12 cm · **2.** *macranthus*
3. Ligule brown, delicate and membranous, at least 1.5 cm long· · · · · · · **6.** sp. A
– Ligule green, leathery, less than 1 cm long · 4
4. Each bract subtending 2 complete flowers · · · · · · · · · · · · · · · · · · **3.** *afer*
– Each bract subtending a single complete flower· 5
5. Ligule with an tangled cottony margin and ring of hairs at junction of petiole; midrib hairy beneath · **4.** *sarmentosus*
– Ligule and midrib glabrous beneath· **5.** *subbiflorus*

1. **Costus spectabilis** (Fenzl) K. Schum. in Bot. Jahrb. Syst. **15**: 422 (1892); in Engler, Pflanzenr. IV. **46**: 421 (1904). —Hutchinson & Dalziel in F.W.T.A. **2**: 332 (1936). —Lock in Kew Bull. **39**: 841 (1984); in F.T.E.A., Zingiberaceae: 4, fig.1 (1985). Type: Sudan, Fassogli, *Boriani* s.n. (?W† holotype). FIGURE 13.4.**19**.
    Cadalvena spectabilis Fenzl in Sitzungsber. Kaiserl. Akad. Wiss., Math.-Naturwiss. Cl., Abt. 1 **51**: 140 (1865). —Baker in F.T.A. **7**: 297 (1898).

Costus pistiifolius K. Schum. in Bot. Jahrb. Syst. **15**: 424 (1892); in Engler, Pflanzenr. IV. **46**: 423 (1904). Type: Angola, Malange, *Teuscz* 315 (B† holotype).

Cadalvena pistiifolia (K. Schum.) Baker in F.T.A. **7**: 297 (1898).

Perennial herb. Rhizome 2–3 mm in diameter when dry; scales, brown, scarious, appressed-pilose in their lower half. Leafy stem borne laterally on rhizome, erect, bearing several reduced leaves and 4 (occasionally 3) broadly spathulate normal leaves, imbricate, arranged like a cross and lying flat on ground. Lamina of normal leaves 4–17 cm broad and long, obtuse to retuse at apex, cuneate at base, glabrous above, appressed-pubescent beneath particularly on midrib, green or yellowish-green with a pink or sometimes brownish or reddish margin. Inflorescence terminal, arising at centre of leaf-rosette, 6–12-flowered, usually only one flower open at any one time; floral bracts delicate, triangular, 1.7 × 0.8 cm. Calyx 2–3 cm long, tubular, membranaceous. Corolla tube 2.5–4.5 cm long; petals 3, yellow, rarely bright orange, narrowly ovate, acuminate, glabrous, or pubescent at apex, 2–4.5 cm long; labellum large, delicate, yellow, 2–5 cm long. Fertile stamen 4.5–6 cm long, ligulate, petaloid, narrowed above anther; anther-thecae 5–7 mm long. Fruits subterranean, poorly known; seeds unknown.

Zambia. N: Mporokoso Dist., Mporokoso to Mkupa, fl. 25.x.1949, *Bullock* 1364 (K). W: Mwinilunga Dist., 16 km from Mwinilunga to Solwesi, fl. 22.xi.1972, *Strid* 2659 (K). C: Lusaka Dist., 8 km E of Lusaka, 1250 m, fl. 16.xii.1955, *King* 247 (K). E: Chadiza Dist., Chadiza, 850 m, fl. 25.xi.1958, *Robson* 682 (BM, K, LISC, SRGH). S: Kalomo Dist., Sichifulo, Siakaunda Hill, fl. 21.xi.1961, *Mitchell* 12/22 (SRGH). **Zimbabwe**. N: Hurungwe Dist., Mrukwe R., Kanyonga stream, 910 m, fl. 18.xi.1953, *Wild* 4179 (SRGH). E: Mutare Dist., Burma Valley, 3.x.1957, *Chase* 6718 (K, SRGH). **Malawi**. N: Nkhata Bay Dist., Kawalazi Estate, 640 m, fl. 17.xii.1985, *la Croix* 757 (K). C: Dedza Dist., Mua Escarpment, fl. 13.xi.1967, *Salubeni* 885 (SRGH). **Mozambique**.

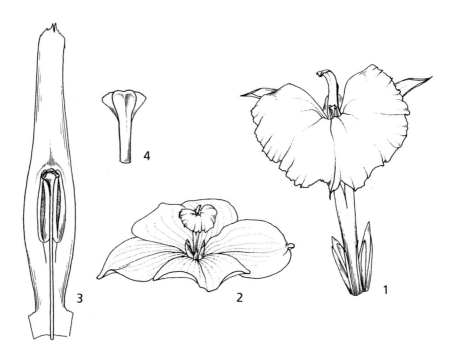

Fig. 13.4.**19**. COSTUS SPECTABILIS. 1, habit (× ¹/₄); 2, flower (× ²/₃); 3, stamen and style (× 2); 4, stigma, rear view (× 6). 1–4 from *Dale* s.n. Drawn by Christine Grey-Wilson. Adapted from Flora of Tropical East Africa.

N: Maúa Dist., Mecopo Mt, fl. 15.x.1942, *Mendonça* 857 (LISC). MS: Barué Dist., near Catandica (Vila Gouveia), fl. 1.xi.1941, *Torre* 3738 (LISC, LMU, MO).

Widespread in the drier parts of seasonal tropical Africa from Sierra Leone to Ethiopia, and southwards to S Angola and Zimbabwe. In *Brachystegia*, *Acacia–Combretum* or *Isoberlinia* woodland, or mixed deciduous woodland and grassland, sometimes in shallow soils over granite; 150–1640 m.

Conservation notes: Widespread species in a widespread habitat; not threatened.

2. **Costus macranthus** K. Schum. in Bot. Jahrb. Syst. **30**: 279, t.7–8 (1901); in Engler, Pflanzenr. IV. **46**: 421 (1904). —Lock in Kew Bull. **39**: 841 (1984); in F.T.E.A., Zingiberaceae: 6 (1985). Type: Tanzania, Rungwe Dist., Kondeland, 1899, *Goetze* 1488 (B† holotype, BM, BR, EA).

   Cadalvena spectabilis sensu C.H. Wright in Bot. Mag. **131**: t.7992 (1905), non Fenzl.

Perennial herb up to 20 cm tall. Rhizome 4–7 mm in diameter when dry; scales brown, scarious, appressed-pilose on lower half. Leafy stem borne laterally on rhizome, erect, with a few reduced leaves and (3)4 imbricate, cross-shaped normal leaves lying flat on ground. Lamina of normal leaves obovate, sometimes almost rounded, 7–19(30) × (5)7–22 cm, apex obtuse, with broad cuneate base, glabrous above, appressed-pubescent beneath, green with pink or reddish margin. Inflorescence developing with the leaves or just before, but flowering continuing until leaves are almost fully developed, terminal in centre of leaf rosette, 6–10-flowered, usually only one flower open at a time; floral bracts delicate, elliptic, acute, c.1.7 × 0.5 cm. Calyx spathaceous, delicate, (3.5)4.5–5 cm long, glabrous. Corolla-tube 7–9 cm long; petals narrowly elliptic, acuminate, 6–8 cm long, pubescent near apex, yellow; labellum large, delicate, yellow, 8–11 cm long. Fertile stamen ligulate, petaloid, narrowed above anthers; anthers 9–13 mm long. Fruit subterranean, otherwise unknown; seeds unknown.

Zambia. N: Mbala Dist., between Chenga and Nakonde R., 85 km from Nakonde, 1500 m, fl. 29.xi.1967, *Richards* 22787 (K). E: Petauke Dist., Sasare, 9.xii.1958, *Robson* 872 (BM, K, LISC). **Zimbabwe**. N: Hurungwe Dist., Mrukwe R., Kanyonga stream, 910 m, fl. 17.xi.1953, *Wild* 4178 (K, SRGH). E: Mutare Dist., Bvumba (Vumba) Mts, 'Falling Waters', 1130 m, fl. 6.xi.1955, *Chase* 5849 (BM, K, SRGH). **Malawi**. N: Karonga Dist., Karonga, Chaminade Sec. School, 550 m, fl. 30.xii.1976, *Pawek* 12101 (K, MAL, UC, SRGH). C: Kasungu Nat. Park, Kasungu, 3.xii.1970, *Hall-Martin* 1028 (K, SRGH). S: Mangochi Dist., 40 km SW of Mangochi (Fort Johnston), fl. 24.xi.1956, *Turner & Shantz* 4150 (SRGH). **Mozambique**. N: Mueda Dist., 47 km from Mueda to Nairoto (Nantulo), 300 m, fl. 4.i.1964, *Torre & Paiva* 9851 (BR, COI, K, LISC, LMA, MO). Z: Mocuba Dist., 67 km from Mocuba to Milange, fl. 11.xi.1942, *Mendonça* 1368 BR, COI, K, LISC, LMU). MS: Gondola Dist., Pindanganga Forest, fl. 17.x.1945, *Simão* 593 (LISC).

Also in Tanzania. Miombo woodland, open bush, and grassland, sometimes among rocks or in sandy soil; up to 1640 m.

Conservation notes: Widespread species in a widespread habitat; not threatened.

3. **Costus afer** Ker Gawl. in Bot. Reg. **8**: t. 683 (1823). —Baker in F.T.A. **7**: 299 (1898). —Schumann in Engler, Pflanzenr. IV. **46**: 392 (1904). —Lock in Kew Bull. **39**: 842 (1984); in F.T.E.A., Zingiberaceae: 9 (1985). Type: Illustration of cultivated plant from Sierra Leone, t.683 in Bot. Reg. **8** (1823).

   Costus pterometra K. Schum. in Engler, Pflanzenr. IV. **46**: 394 (1904). Type: Congo/Sudan border, 10.ii.1870, *Schweinfurth* ser.3, 204 (K lectotype, designated by Lock 1984, PRE).

Perennial herb with erect leafy stems 2–4 m tall. Leaves elliptic to obovate, 15–35 × 3.5–9.5 cm, apex acuminate, narrowed below to rounded or subcordate base, margin sparsely ascending-ciliate, lamina otherwise usually glabrous but sometimes sparsely appressed-pubescent beneath; petiole 4–12 mm long; ligule coriaceous, green in life, 4–8 mm long, glabrous; sheaths lightly striate when dry, in life green, often with purple blotches. Inflorescence terminal on leafy shoots; floral bracts oblong, convex, truncate to rounded at apex, c.3.5 × 3 cm, upper ones rather smaller, green with purple markings, each subtending 2 fully developed flowers; bracteoles boat-shaped, with thickened rigid keel, pale green with pink markings and thin pink scarious margin, c. 2.5 × 0.8 cm. Calyx funnel-shaped, 1.7–2 cm long, with 3 triangular teeth 5 mm long, with a pink scarious margin; calyx teeth project 2–3 mm beyond bracts after anthesis. Corolla tube c. 2 cm long, hairy inside; petals white, semi-transparent, oblong-ovate, 3–4 × 1.2–1.4 cm, hooded at apex; labellum broadly triangular, funnel-shaped, c.2.5 cm long and wide, white, sometimes tinged pink at tip, a rich orange-yellow central line extending to base of the tube, with undulate margin. Stamen ovate, c.3 × 1.2 cm, white; anthers 7–8 mm long. Fruit c.1 × 0.6 cm long, obovoid; seeds 2 × 1 mm, irregularly oblong, black.

Zimbabwe. E: Chimanimani Dist., Haroni-Makurupini Forest, by river, 400 m, fl. 3.xii.1964, *Wild, Goldsmith & Müller* 6625 (K, SRGH). **Malawi**. N: Nkhata Bay Dist., Nkhwadzi (Nkwazi) Forest, 14.5 km S of Nkata Bay junction, 600 m, fl.& fr. 22.ii.1976, *Pawek* 10866 (K, MAL, MO, SRGH, UC).

Tropical Africa from Sierra Leone east to Ethiopia, and south to the Flora area. Moist places in forest, often by streams and rivers, forest margins; 300–600 m.

Conservation notes: Widespread species; not threatened.

4. **Costus sarmentosus** Bojer in Ann. Sci. Nat. Bot., sér.2 **4**: 262, t.8 (1835). —Baker in F.T.A. **7**: 299 (1898). —Schumann in Engler, Pflanzenr. IV. **46**: 394 (1904). —Lock in F.T.E.A., Zingiberaceae: 7 (1985). Type: Drawing of plant from Zanzibar Is., t.8 in Ann. Sci. Nat. Bot., sér.2 **4** (1835).

Perennial herb, erect leafy stems 1.5–3.5 m tall. Leaves elliptic to ovate, 14–31 × 4–10 cm, apex acuminate, narrowed below to rounded or subcordate base, appressed-pubescent on midrib beneath and ascending-pilose on margins, otherwise glabrous; petiole 3–10 mm long; ligule coriaceous, 5–12 mm long, with marginal fringe of tangled hairs, sometimes with a line of hairs partially surrounding stem at ligule base; sheaths striate, glabrous. Inflorescence usually terminal on leafy shoots, sometimes borne on separate leafless bracteate shoots; floral bracts broadly ovate, strongly convex, rounded to truncate at apex, 2–2.5 × 1–1.5 cm, each subtending a single flower; bracteole boat-shaped, coriaceous, 1.8–2 × 0.5–0.8 cm. Calyx tubular, 1.4–1.5 cm long, lobes c.3 mm long, incurved. Corolla tube 1.4–1.7 cm long, hairy inside; petals white, oblong to obovate, 3–3.2 × 1–1.2 cm, hooded, sometimes apiculate or retuse at apex; labellum broadly triangular, funnel-shaped, c.4 cm long and wide, white, flushed pink at apex, with yellow patch in centre. Stamen elliptic, 2–2.5 × 1–1.2 cm; anthers c.9 mm long. Fruit obovoid, c.12 × 9 mm; seeds oblong, c.2 × 1 mm, dark grey, finely striate.

Malawi. N: Nkhata Bay Dist., Nkhwadzi (Mkuwadzi) Hill Forest, fl.& fr. 13.iv.1960, *Adlard* 333 (K, SRGH).

Also in Kenya and Tanzania. Woodland, evergreen forest, and forest margins; c.700 m.

Conservation notes: Not threatened globally, but very local in Flora area; perhaps Near Threatened.

5. **Costus subbiflorus** K.Schum. in Engler, Pflanzenr. IV. **46**: 394 (1904). —Lock in F.T.E.A., Zingiberaceae: 8 (1985). Types: Tanzania, Lushoto Dist., Magila, *Volkens* 50 (B† syntype, BM); Amani, *Engler* 715 (B† syntype); Kwa Mburaka–Kisula, *Buchwald* 360 (B† syntype); Pamora, *Liebusch* 17 (B† syntype). FIGURE 13.4.**20**.

Perennial herb with erect leafy stems 2–5 m tall. Leaves elliptic, 20–25 × 4.5–7 cm, apex acuminate, narrowed below to cuneate to subcordate base, glabrous except for line of ascending marginal hairs; petiole 3–5 mm long; ligule coriaceous, 5–7 mm long, glabrous; sheaths striate when dry, glabrous. Inflorescence terminal on leafy shoots, sometimes

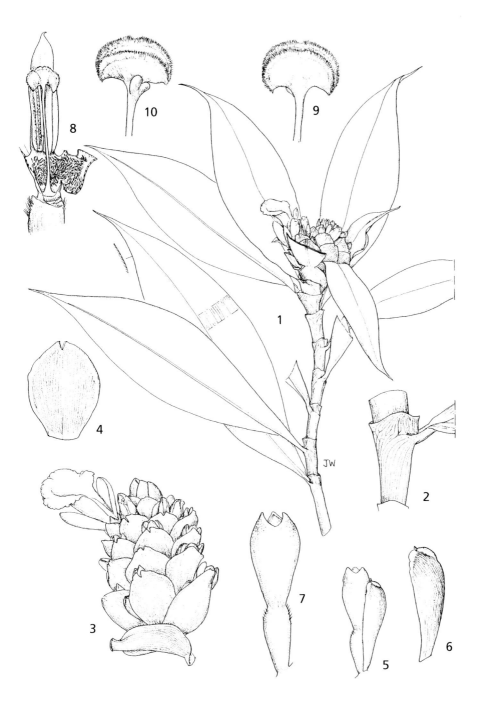

Fig. 13.4.**20**. COSTUS SUBBIFLORUS. 1, habit (× $^1/_3$); 2, leaf base (× 1); 3, inflorescence (× $^2/_3$); 4, bract (× 1); 5, calyx with ovary and bracteole (× 1); 6, bracteole (× $1^1/_2$); 7, calyx and ovary (× $1^1/_2$); 8, stamen and anthers (× 3); 9, stigma, anterior (× 8); 10, stigma, posterior (× 8). All from *Torre & Correia* 16226. Drawn by Juliet Williamson.

terminating separate basal shoots; floral bracts broadly obovate, strongly convex, apiculate, 2.5–3 × 2 cm, each subtending a single flower; bracteoles boat-shaped, keeled, c.2.5 × 1 cm. Calyx funnel-shaped, not exceeding bracts, c.1.7 cm long, with 3 obtuse teeth c.4 mm long. Corolla tube c.1.3 cm long, hairy inside; petals white, oblong, apiculate, c.3.5 × 1.5 cm, white with central yellow mark. Stamen narrowly ovate, c.3 × 1 cm; anthers c.9 mm long. Fruit obovoid, c.12 mm long; seeds oblong, c.2 mm long, grey, finely striate.

Zimbabwe. E: Chimanimani Dist., Rusitu (Lusitu) R. c.8 km W of Haroni junction, c.300 m, fl. 20.ix.1960, *Rutherford-Smith* 125 (K, LISC). **Malawi**. N: Nkata Bay Dist., Viphya Plateau, fl. 6.iv.1962, *Chapman* 1656 (K). **Mozambique**. Z: Maganja da Costa Dist., 45 km from Vila da Maganja on road to Mocuba, 40 m, fl.& fr. 8.xi.1966, *Torre & Correia* 14505 (BR, COI, K, LISC, LMU, MA, MAL, MO, SRGH). MS: Sussundenga Dist., c.20 km W of Dombe, SE end of Chimanimani Mts, 350 m, fl. 24.iv.1974, *Pope & Müller* 1279 (SRGH).

Also in Tanzania. Evergreen forest and along streams; 40–700 m.

Conservation notes: Widespread species; not threatened.

6. **Costus** sp. **A.**

Perennial erect herb to 2.5 m. Leaves oblanceolate to obovate, 14–26 × 4–8.7 cm, glabrous, apex acuminate, narrowed below to base gradually and finally ± abruptly contracted into 6–10 mm long petiole. Ligule bilobed, membranous, brown, 1.5–4.5 cm long. Inflorescence terminal on leafy shoot; floral bracts 2.5 × 1.5 cm, acute or obtuse, foliaceous at apex, each subtending a single flower. Calyx funnel-shaped, c.1.5 cm long, 3-dentate, split partially on one side. Petals c.5 cm long, pink or white-pinkish/red, yellow inside at base. Labellum c.6 cm long, pinkish. Anthers c.7 mm long. Fruit immature 1–1.3 cm long.

Zambia. N: Mansa Dist., Mapula Valley, 45 km from Kwamabwa on road to Katotoma, fl. 10.iv.1961, *Angus* 2795 (K, SRGH). W: Mwinilunga Dist., Lisombo R., fr. 14.vi.1963, *Edwards* 776 (K, SRGH). C: Lusaka Dist., Chilanga, Mt Makulu, fl. i.1971, *Anton-Smith* in *SRGH* 213,252 (SRGH).

Not known elsewhere. In riparian forest and moist places in forest; c.1350 m.

Conservation notes: Apparently endemic to Zambia, but moderately widespread; probably Near Threatened.

This species, with its long brown membranous ligules, seems closest to *C. fissiligulatus* Gagnep. from Gabon and Cameroon, but that species has hairy undersides to the leaves and is a smaller plant. *C. dewevrei* De Wild. & T. Durand is also similar but is very hairy. More material and field observations are needed.

ZINGIBERACEAE

by J.M. Lock

Perennial herbs, usually with creeping rhizomes, terrestrial but occasionally epiphytic (not in the Flora area); sterile and fertile shoots often separate, usually unbranched. Vegetative parts with oil glands, aromatic. Leaves solitary at nodes or crowded at stem base, alternate, in 2 rows, consisting of sheath, ligule and lamina; sheaths encircling stem, sometimes forming a pseudostem, margins free; ligule adaxial, often bilobed; lamina usually narrowly elliptic, narrowing into a short pseudopetiole, with a strong median nerve and parallel laterals diverging from the midrib at c.45°; all parts glabrous or hairy with simple unicellular hairs. Inflorescences racemose (spikes, racemes or panicles), bracts and bracteoles (if present) usually large and conspicuous, sometimes coloured. Flowers hermaphrodite, strongly zygomorphic, trimerous. Calyx tubular or spathaceous, often 3-lobed at apex, valvate. Petals 3, equal or unequal, fused at base with androecium to form a perianth-tube, overlapping in bud. Androecium composed of a single often petaloid stamen with anther thecae dehiscing by longitudinal slits, a large petaloid variously-lobed labellum formed from 2 fused staminodes (filamentous outgrowths at the stamen base are sometimes claimed to represent other staminodes). Ovary inferior, trilocular, placentation axile, usually with numerous ovules; style and stigma one. Nectaries of various forms occur either beside the style-base (those that are paired and elongated have been called stylodia) or embedded in top of ovary. Fruit a berry or capsule, often large and with thick fleshy walls, sometimes subterranean. Seeds usually smooth and hard, with abundant endosperm, surrounded by a fimbriate aril.

A pantropical family with c.50 genera and perhaps 900 species, mostly in areas of high rainfall. Most abundant and diverse in SE Asia.

The series of papers on the family by Burtt and Smith (Notes Roy. Bot. Gard. Edinb. **31**: 155–227, 1972) should be read by all would-be students of the family. Their key to genera (pp. 171–176) has, however, been superseded, but the new version remains unpublished.

Several members of the family are cultivated for their spicy rhizomes or seeds.

Zingiber officinale Roscoe is widely cultivated for its rhizomes (Root Ginger). The leaves, borne in 2 rows on an elongated stem, are very narrowly elliptic; the inflorescence (rarely produced) bears a bracteate head of dark purple-brown flowers. The smell of the cut rhizome is also distinctive. Zimbabwe N: Mazoe, 'cult. ex Amani', 17.iv.1950, *Pollitt* in SRGH 27435 (SRGH).

Zingiber zerumbet (L.) Sm. is also cultivated, usually as an ornamental plant but sometimes (in error) as "ginger". It differs from *Z. officinale* in its larger size, broader leaves, larger ligule, and flowers with a pale yellow, not deep purple-brown, labellum.

Curcuma longa L. is also cultivated for its rhizomes (Turmeric) (e.g. Zimbabwe C: Harare, *Biegel* 5566 (SRGH)). The leaves form a clump and are long-petiolate; the inflorescence is borne at the end of a long peduncle and is usually surmounted by a group of coloured bracts.

Elettaria cardamomum (L.) Maton is the source of the cardamoms of commerce (the seed pods).

Species that are commonly cultivated as ornamentals include:

Etlingera elatior (Jack) R.M. Sm. (*Nicolaia elatior* (Jack) Horan., *Phaeomeria magnifica* (Roscoe) K. Schum.; Porcelain Rose) has cone-like heads of waxy-pink bracts bearing small red and yellow flowers in their axils and is often cultivated (e.g. Maputo (Lourenço Marques), Jardim Vasco da Gama, 20.xii.1973, *Balsinhas* 2598 (SRGH)).

Species of *Hedychium*, particularly *H. coronarium* Koenig (Ginger Lily), are often cultivated (e.g. Zimbabwe C: Harare, *Biegel* 2902 (SRGH); Malawi S: Zomba Botanic Garden, *Salubeni* 2653 (SRGH)). *Hedychium coronarium* has large white sweet-scented flowers borne in loose inflorescences at the apex of a leafy stem. Other species have yellow or orange flowers.

Of the cultivated species of *Alpinia*, *A. purpurata* (Vieill.) K. Schum. has numerous persistent purple-red bracts around the inflorescence. *Alpinia speciosa* (Wendl.) K. Schum. (Shell Ginger) has white flowers marked with red and yellow at the apex of the leafy stems (Zimbabwe C: cult. Harare, *Whellan* in SRGH 50608 (SRGH)). The source of the spice Galangal, *Alpinia galanga* (L.) Willd., has also been recorded as cultivated in Malawi (S: Zomba, *Salubeni* 2387 (SRGH)).

1. Inflorescence a many-flowered panicle with an elongated axis; flowers, including pedicel, less than 2.5 cm long · **1. Renealmia**
– Inflorescence condensed, without an elongated axis; if axis elongated, then each bract subtending only one flower which, with pedicel, is more than 2.5 cm long ·2

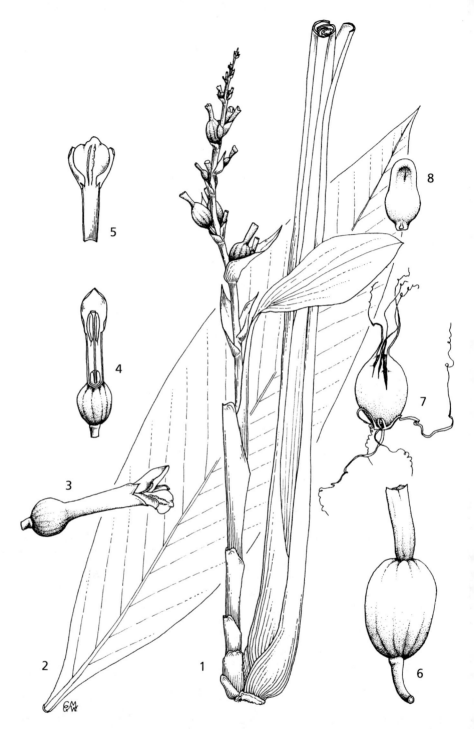

Fig. 13.4.**21**. RENEALMIA ENGLERI. 1, base of plant with inflorescence (× $^2/_3$); 2, leaf blade (× $^2/_3$); 3, flower (× 2); 4,5, longitudinal section of flower (× 2); 6, fruit (× 2); 7,8, seed with and without aril (× 6). All from *Drummond & Hemsley* 1722. Drawn by Christine Grey-Wilson. From Flora of Tropical East Africa.

2. Apex of stamen petaloid, at least twice as long as basal anther; small plants less than 1 m tall, often flowering before leaves are fully developed; leaf sheaths of some species forming a pseudostem, but not at flowering time · · **2. Siphonochilus**

– Apex of stamen 2–3-lobed or truncate, shorter than subterminal anther; large plants with leafy shoots 1 m or more tall, usually fully developed at flowering time; leaf sheaths always forming a distinct pseudostem · · · · · · **3. Aframomum**

1. **RENEALMIA** L.f.

Renealmia L.f., Suppl. Pl.: 7, 79 (1782), nom. conserv.

Perennial herbs, with leafy stems in clumps from short rhizomes. Leaves in 2 rows with long sheathing bases forming a pseudostem, narrowly ovate to obovate, attenuate at base into a pseudopetiole; ligule very short and stiff; sheaths furrowed, sometimes pitted or reticulate. Inflorescences paniculate, arising from base of leafy shoots or sometimes (not in Flora area) terminating them. Inflorescence-axis bearing oblong to ovate bracts, subtending small cincinni of 1–8 or more flowers, each subtended by a tubular bracteole. Calyx tubular, trifid at the apex, persistent in fruit. Corolla 3-lobed, attached below to androecial tube. Androecium composed of a 3-lobed labellum and a single subsessile stamen. Ovary with 1 to many seeds (9 to many in the Flora area); style filiform, with a bilobed stigma; stylodia (nectaries) forming a solid mass around the style base. Fruit a fleshy capsule, may be tardily dehiscent. Seeds shiny, striate, angular, surrounded by a fimbriate aril.

About 55 species in tropical America and 10–20 species in Africa, with some recent revisionary work by Dhetchuvi, although his work remains unpublished. The American species were revised by Maas (Fl. Neotropica Monogr. **18**, 1977).

1. **Renealmia engleri** K. Schum. in Engler, Pflanzenr. IV. **46**: 293 (1904). —Lock in F.T.E.A., Zingiberaceae: 11 (1985). Types: Tanzania, Lushoto Dist., Ngwelo, *Heinsen* 64 (B† syntype, K) & Amani, *Engler* (B† syntype). FIGURE 13.4.**21**.

Perennial herb with leafy stems up to 2.5 m high, forming a dense clump. Leaves narrowly obovate, 30–55 × 5.5–9 cm, abruptly caudate-acuminate, attenuate at base with a petiole 2–5 cm long, glabrous; ligule very short, rounded; sheaths furrowed, sparsely reticulate, minutely puberulous in furrows. Inflorescence arising at base of leafy shoot, 20–30 cm or more tall; peduncle bracts 6–12 × 1.5–2 cm, papery, ± persistent in fruit, shortly pubescent particularly towards apex; floral bracts oblong, rounded, 1.5 × 0.5 cm, glabrescent; bracteoles tubular, pubescent. Calyx pale orange-brown, tubular, 3-toothed at apex, pubescent, particularly towards apex. Petals whitish, translucent, oblong, apex rounded, 3.5 × 2 mm. Labellum 12 mm long, the free portion 3-lobed, 7.5 mm wide × 6 mm long. Free filament 3.5 mm long, apex rounded; anther 3.5 mm long, completely dehiscent. Ovary 3 mm long, glabrous; style filiform. Fruit subglobose to cylindrical, 6–12 × 5–7 mm, glossy black (*Fanshawe* 2984), crowned by a persistent orange calyx 6–7 mm long, with 3 appressed-pubescent teeth at apex. Seeds 9–12 in each fruit, c.3 × 2.5 mm, chestnut-brown, finely striate, shiny.

Zambia. N: Kawambwa Dist., Kawambwa, fl.& fr. 30.i.1957, *Fanshawe* 2984 (K, NDO); Kawambwa, fr. 28.v.1962, *Lawton* 875 (K).
Also in Kenya and Tanzania. Swamp forest; c.1200 m.
Conservation notes: In the Flora area known only from two collections at a single locality in a habitat that is threatened by clearance for agriculture. Probably Vulnerable in the Flora area, but Near Threatened on continental scale.
Dhetchuvi has determined the cited material as this species but the distinctions between *R. engleri* and *R. congolana* De Wild. & T. Durand remain somewhat unclear. The taxonomy of the African species is still unsatisfactory.

2. **SIPHONOCHILUS** J.M. Wood & Franks

Siphonochilus J.M. Wood & Franks in Bull. Misc. Inform., Kew **1911**: 274 (1911);
in Wood, Natal Plants **6**: t.560, 561 (1911). —Burtt in Notes Roy. Bot. Gard. Edinb.
40: 369–373 (1982).

Cienkowskia Solms in Schweinfurth, Beitr. Fl. Aeth.: 197 (1867), non Regel & Rach (Index
Semin. Hort. Petrop.: 48, 1858).

Kaempferia L. subgen. *Cienkowskia* K. Schum. in Engler, Pflanzenr. IV. **46**: 67 (1904).

Cienkowskiella Y.K. Kam in Notes Roy. Bot. Gard. Edinb. **38**: 8 (1980).

Perennial herbs from a short swollen rhizome or corm. Roots often tuberous. Leafy
shoots often produced after flowering; lamina narrowly to broadly elliptic, glabrous,
ligule scarious or absent; sheaths furrowed when dry, sometimes forming a
pseudostem. Inflorescences racemose, 2–20-flowered, separate from leafy shoot, often
precocious, on an elongated or very short peduncle, and concealed among basal
bracts and old leaf sheaths. Peduncle with several bracts at base; pedicellate flowers
each subtended by a single floral bract; bracteoles absent. Calyx tubular or obconical,
3-lobed at apex, each lobe often with a subterminal subulate projection. Corolla with
a basal tube and 3 free subequal petals. Androecium fused basally with corolla tube,
comprising a 3-lobed labellum with a usually deeply divided central lobe and a single
stamen with basal anthers and a long petaloid terminal lobe. Ovary subspherical,
sometimes trigonous; style filiform, stigma cup-shaped or 2-lipped, glabrous;
epigynous glands very short. Fruits often ± subterranean except in species with long-
pedunculate inflorescences, generally poorly known. Mature seeds unknown.

About 15 species in savannas and dry forests of tropical Africa; one recently-
described species (*S. bambutiorum* A.D. Poulsen & Lock) from moist forest. The
reasons for regarding the African plants as a genus distinct from the Asian *Kaempferia*
have been fully set out by Kam (1980).

Several species flower precociously and knowledge of their leaves is poor. The
frequent absence of mature leaves from collections, the extreme delicacy of the
flowers and the difficulty of making satisfactory dissections from dried material
means that there is still much work to be done; this account should be regarded
as preliminary. In the author's experience, material dug up in the field will often
establish well in cultivation, and local botanists could contribute valuable data by
growing these species and collecting mature leaves. Extra flowers, dried
separately, and dissections, particularly ones showing the form of the stamen, are
very useful.

1. Flowers on elongated peduncle · 2
 – Flowers on very short peduncle concealed by basal bracts · · · · · · · · · · · · · · 4
2. Leaf blade narrowly elliptic, 7–10 times as long as wide · · · · · · · · · · · · · · · · 3
 – Leaf blade broadly elliptic, 2–5 times as long as wide · · · · · · · · · · · · · **1.** *kirkii*
3. Inflorescence 3–7-flowered; labellum deep purple, large; leaves 22–30 × 2–3.5
 cm; N Zambia· **2.** *pleianthus*
 – Inflorescence 4–10-flowered; labellum dull purple or mauve, small; leaves
 smaller; coastal Mozambique · **3.** *kilimanensis*
4. Labellum white, purple or mauve, more than 3.5 cm long; corolla lobes more
 than 2.5 cm long; anther-crest ovate, obovate or elliptic, 5 mm wide or more · 5
 – Labellum dull mauve, up to 3.5 cm long; corolla lobes 2–2.5 cm long; anther-
 crest parallel-sided, c.2 mm wide · **6.** *parvus*
5. Flowers white with corolla tube at least $^2/_3$ of labellum length, or blue with the
 corolla tube longer than free part of labellum · 6
 – Flowers mauve to purple (very occasionally white); corolla tube shorter· · · · · 8

6. Flowers blue; labellum almost circular in outline; leaves fully developed at anthesis ·· **9.** *brachystemon*
– Flowers white; labellum lobed; leaves hardly developed at anthesis, or absent ·· 7
7. Ligule short and thick; root tubers subspherical; sterile apex of stamen ovate, 8–12 × 6–8 mm; perianth segments often spirally reflexed after anthesis; leaves usually developing at anthesis ····························· **8.** *longitubus*
– Ligule longer, membranous; root tubers narrowly elliptic; sterile apex of stamen obovate, c.12 × 5–6 mm; perianth segments not usually reflexed after anthesis; leaves usually absent at anthesis ···························· **5.** *rhodesicus*
8. Flowers pale mauve-pink to blue-purple (very occasionally white); dry perianth with only scattered minute brown dots (oil glands); at least young leaves often present with flowers ································· **4.** *aethiopicus*
– Flowers mauve-pink; dry perianth with numerous minute brown dots (oil glands); leaves usually completely absent at flowering ········· **7.** *puncticulatus*

1. **Siphonochilus kirkii** (Hook. f.) B.L. Burtt in Notes Roy. Bot. Gard. Edinb. **40**: 372 (1982). —Lock in F.T.E.A., Zingiberaceae: 15, fig.4 (1985). Type: Tanzania coast opposite Zanzibar, plant collected by Kirk and grown at Kew, fl. vii.1880 (K holotype). FIGURE 13.4.**22**.

> *Cienkowskia kirkii* Hook. f. in Bot. Mag. **98**: t.5994 (1872).
> *Kaempferia kirkii* (Hook. f.) Wittm. & Perring in Gartenflora **41**: 57, t.1364 (1892). — Baker in F.T.A. **7**: 294 (1898). —Schumann in Engler, Pflanzenr. IV. **46**: 68 (1904).
> *Kaempferia rosea* Baker in F.T.A. **7**: 295 (1898). —Troupin in Bull. Jard. Bot. État. **25**(4): 266 (1955). Types: Sudan, Jur Ghattas, 28.v.1869, *Schweinfurth* 1946 (K syntype) and numerous other specimens from eastern Africa.
> *Kaempferia carsonii* Baker in F.T.A. **7**: 296 (1898). Type: Malawi, Karonga, v.1891, *Carson* s.n. (K holotype).
> *Kaempferia pallida* De Wild. in Ann. Mus. Congo, sér.4: 20 (1902). Type: Congo, Katanga, xi.1892, *Verdick* 297 (BR).
> *Kaempferia kirkii* var. *elatior* Stapf in Bot. Mag. **134**: t.8188 (1908). Type: Illustration t. 8188 of plant from Zimbabwe (Rhodesia) cultivated in Britain, 17.vi.1907, *Elwes* s.n. (K).
> *Kaempferia montagui* Leighton in S. Afr. Country Life **22**: 57 (1932). Type: Zimbabwe, near Mazoe, i.1932 & 10.iii.1939, *Montagu & Wiese* in NBG 888/21 and NBG 33/26 (K holotype, BOL). A specimen in SRGH (QVM 7886) is labelled "*Kaempferia montagui* Leighton – from Montagu's old garden".
> *Kaempferia decora* Druten in Fl. Pl. Afr. **30**: t.1199 (1955). Type: Mozambique, Garuso, in *Brachystegia* woodland, plant cult. in Pretoria, fl. iii.1953, *Schweickerdt* in PRE 28519 (PRE holotype).
> *Cienkowskiella kirkii* (Hook.f.) K.Y. Kam in Notes Roy. Bot. Gard. Edinb. **38**: 11 (1980).
> *Siphonochilus carsonii* (Baker) Lock in Kew Bull. **39**: 841 (1984); in F.T.E.A., Zingiberaceae: 17 (1985).
> *Siphonochilus decorus* (Druten) Lock in Kew Bull. **54**: 349 (1999).

Perennial herb from a short thick rhizome. Roots with tuberous swellings towards apex, to 2.5 × 0.8 cm. Leaves fully developed or developing at time of flowering; lamina glabrous, ovate to elliptic, acuminate, 17–31 × 5.5–10 cm, gradually tapering at base into false petiole up to 25 cm long including leaf sheath; ligule obsolete; sheaths furrowed when dry; pseudostem very short or absent; lateral veins prominent beneath, 3.5–7 mm apart when dry. Inflorescence arising at base of leafy shoot, 7–15(20)-flowered; peduncle 20–35 cm long, terete, glabrous; bracts 3–4, greenish, oblong, rounded, emarginate, up to 6 × 1.5 cm, floral bracts greenish, oblong to narrowly obovate, obtuse, the lower up to 6.5 × 2 cm, upper very much smaller, to 2.2 × 1.2 cm. Calyx campanulate, 8–15 mm long, not markedly increasing in fruit, shallowly 3-lobed, each with a subterminal subulate projection c.1 mm long. Corolla tube c.8 mm long; petals narrowly obovate to narrowly oblong, acute, 2.2–2.6 cm long, whitish, tinged with green or mauve. Labellum 3-lobed, lateral lobes rhomboid, mauve, c.3 × 2 cm, the median broadly spathulate, emarginate, c.4.5 × 4.5 cm, mauve with a central yellow mark, sometimes with a dark

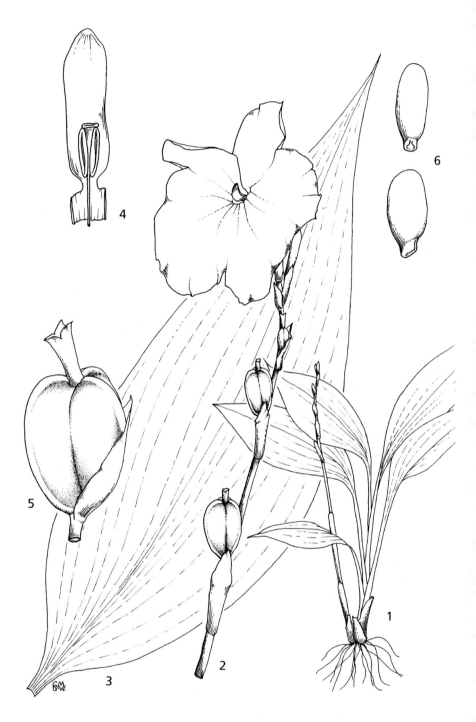

Fig. 13.4.**22**. SIPHONOCHILUS KIRKII. 1, habit (× ¹/₄); 2, inflorescence (× ²/₃); 3, leaf blade (× ²/₃); 4, stamen and style (× 2); 5, fruit (× 2); 6, seed (× 4). All from *Polhill & Paulo* 1330. Drawn by Christine Grey-Wilson. From Flora of Tropical East Africa.

purple mark on each side, or completely yellow. Stamen petaloid, oblong or pandurate, 2.2–2.5 × 0.7–0.9 cm; anther 6–7 cm long. Ovary c.6 mm long, trigonous. Fruit obovoid, trigonous, winged at angles, up to 2.5 cm long; fully mature fruit and seeds not seen.

Caprivi Strip. Katima Mulilo area, 5 km S of Katima Mulilo, 21.xi.1960, *Killick & Leistner* 3016 (SRGH). **Zambia**. B: Sesheke Dist., Masese Forest Station, 2.ii.1975, *Brummitt, Chisumpa & Polhill* 14236 (K). N: Kabwe (Broken Hill), Chibwe Forest (P.F.A.), fl. 5 xii.1960, *Morze* 202 (K). C: Lusaka Dist., 11 km from Great East Road turn-off along road to Feira (c.30 km WSW of Luangwa R. Bridge), 30.xii.1972, *Strid* 2880 (K). E: Chadiza Dist., Chadiza, fl. 25.xii.1958, *Robson* 690 (BM, K, LISC). S: Namwala Dist., Ngoma, Kafue Nat. Park, 6.xii.1960, *Mitchell* 2/34 (SRGH). **Zimbabwe**. N: Gokwe Dist., Mabviriviri, Gokwe to Copper Queen road, 20.xii.1963, *Bingham* 1007 (LISC, SRGH). W: Hwange (Sebungwe) Dist., Dete (Dett) Salt Pan, fl. 7.xii.1956, *Lovemore* 514 (K, LISC, SRGH). E: Chipinge Dist., 5 km NE of confluence of Musirizwi and Busizi rivers, 28.i.1975, *Biegel, Pope & Russell* 4830 (SRGH). **Malawi**. N: Nkhata Bay Dist., Vipya plateau, 37 km SW of Mzuzu, Lwafwa Drift, 23.xii.1975, *Pawek* 10435 (K, SRGH). C: Kasungu Dist., Kasungu Nat. Park, 3.x.1970, *Hall-Martin* 1029 (SRGH). S: Mangochi Dist., Monkey Bay to Mangochi road, 11.i.1980, *Masiye, Salubeni & Tawakali* 230 (MAL, SRGH). **Mozambique**. N: Mandimba Dist., Mandimba, 6.xi.1941, *Hornby* 3457 (K). Z: Mopeia Dist., 85 km from Morrumbala on road to Quelimane, 8.xii.1971, *Pope & Müller* 552 (K, LISC, SRGH). T: Angonia Dist., Ulónguè, near Monte Caponda, 16.xii.1980, *Macuácua* LM 1445 (K, LISC, LMU). MS: Gorongosa Dist., 30 km from entrance to Gorongosa Nat. Park (Parque Nacional de Caça) towards Gorongosa (Vila Paiva de Andrade), 12.xi.1963, *Torre & Paiva* 9212 (LISC, LMA, LMU).

Also in Sudan, Uganda, Kenya, Tanzania and Congo. Deciduous woodland, wooded grassland, seasonal and riverine forest and scrub; 100–1350 m.

Conservation notes: Never a very abundant species but widespread; not threatened.

The variable *Siphonochilus kirkii* may be a species in the process of splitting into more than one taxon. In F.T.E.A. (Lock 1985) I treated *S. carsonii* as a separate species, but having now seen a wider range of specimens and re-examined the type, I have decided to treat it as synonymous with *S. kirkii*. The differences between the two are of degree only, with plenty of intermediates. Plants treated as *S. carsonii* in F.T.E.A. occur in mixed populations in which there are both mauve- and yellow-flowered individuals; the difference is not as clear-cut as might at first appear.

Plants formerly named as *Siphonochilus decorus* differ only in flower colour and phenology and are the commonest form in the higher rainfall areas of S Malawi, E Zimbabwe and W Mozambique. Plants with different flower colours occur together – *Torre & Paiva* 9652 and 9653 collected from the same site on the same day in Mozambique N: have flowers described respectively as 'amarelas' [yellow] and 'esbranquiçadas' [whitish].

Siphonochilus pleianthus and *S. kilimanensis* are also close to *S. kirkii*, but are here treated as distinct taxa; they are more clearly differentiated from the variable 'core' of material that comprises *S. kirkii* and have distinct distributions.

2. **Siphonochilus pleianthus** (K. Schum.) Lock, comb. nov. Type: Angola, Cuango, Kitamba, 8.xii, *Buchner* 694 (?B† holotype).

    Kaempferia pleiantha K. Schum. in Bot. Jahrb. Syst. **15**: 425 (1892). —Troupin in Bull. Jard. Bot. État. **25**: 266 (1955).

Perennial herb from underground rootstock. Roots tuberous (*Fanshawe* 2985). Leaves forming a loose rosette, without a pseudostem, usually present or developing at anthesis. Lowest leaves reduced to bracts, 5–16 × 1.2–1.6 cm; foliage leaves 2–4, narrowly elliptic to narrowly ovate, 22–30 × 2–3.5 cm, glabrous, glaucous in life (*Fanshawe* 2985), apex narrowly acute, base gradually narrowing into a broad flattened sheath, drying brownish with darker upper surface and darker brownish margins; midrib c.1 mm broad beneath, pale and clear, lateral veins rather indistinct beneath, 1–1.5 mm apart. Inflorescence arising beside the leafy shoot, (3)4–7-flowered, 35–50 cm tall; sterile bracts oblong, 4–11.5 × 1.5 cm, uppermost the largest, apex broadly obtuse; floral bracts brown, membranous, 2.5–3.5 cm, apiculate, sheathing, 1–1.5 cm long, split to 0.5–0.8 cm. Perianth segments 3, fused into a tube at the base, free lobes narrowly oblong, 3.5–4.5 × 0.6–0.8 cm, laterals acute at apex, the median acuminate. Labellum up to 9 cm long, including tube, the median lobe suborbicular, c.3.5 × 6 cm, purple with orange central markings, lateral lobes broadly obovate, 3.5–4.5 × 3–4 cm, rich purple. Apical lobe of stamen ligulate, tapering above to acute or truncate apex. Anthers, ovary, fruit and seeds not seen.

Zambia. N: Kawambwa, fl. 30.i.1957, *Fanshawe* 2985 (K); Mporokoso Dist., Kabwe Plain, Mweru-Wantipa, salt pans, fl. 14.xii.1960, *Richards* 13695 (K).

Also in Congo and Angola. (Seasonally?) moist grasslands; 900–1200 m.

Conservation notes: Apparently not uncommon where it occurs but its extent of occurrence is small; Near Threatened may be appropriate.

Siphonochilus sp. A of Lock (F.T.E.A., Zingiberaceae: 18, 1985) is certainly part of the same complex, but the flowers there are described as 'white with blood-red markings', and the leaves appear thinner-textured. The taxon has been collected again in the same area since the publication of F.T.E.A. (*Bidgood, Mbago & Vollesen* 2697 (K)).

3. **Siphonochilus kilimanensis** (Gagnep.) B.L. Burtt in Notes Roy. Bot. Gard. Edinb.
    40: 372 (1982). Type: Mozambique, Campo, 22.xii.1904, *Le Testu* 564 (P holotype, BR, LISC).

    Kaempferia kilimanensis Gagnep. in Bull. Soc. Bot. France **53**: 352 (July 1906).
    Kaempferia ceciliae N.E.Br. in Bull. Misc. Inform., Kew **1906**: 169 (31 July 1906). Type: Mozambique, Dondo, xii.1899, *Cecil* 248 (K).

Perennial herb from a short thick rhizome. Root tubers present? Leaves fully developed or almost so at time of flowering; lamina glabrous, very narrowly elliptic, acute, 12–22 × 1.2–2.2 cm, gradually tapering at base into a false petiole to 11 cm long, including leaf sheath; ligule obsolete; sheaths weakly furrowed when dry; pseudostem absent; lateral veins visible but not prominent beneath, 1–2 mm apart when dry. Inflorescence arising at base of leafy shoot, 4–10-flowered; peduncle 19–38 cm long, terete, glabrous; bracts 1–3, brownish, narrowly oblong, rounded, apiculate, up to 10 × 1 cm, floral bracts brownish, oblong to narrowly obovate, obtuse, lower bracts up to 4.5 × 1 cm, upper very much smaller, to 2 × 1 cm. Calyx campanulate, 5–6 mm long, not markedly increasing in fruit, shallowly lobed to undulate. Corolla tube c.5 mm long; petals elliptic to ovate, acute, 1.8–2.2 × 1 cm. Labellum 3-lobed, 4.5–5.5 cm long, lateral lobes rhomboid, violet, median lobes broadly spathulate, emarginate, violet. Anthers c.5 mm long; sterile anther crest oblong, c.1 cm long. Ovary c.5 mm long, trigonous. Mature fruit and seeds not seen.

Mozambique. Z: Maganja da Costa Dist., 5 km from crossing of Namacurra–Maganja and Macuze roads, fl. 29.i.1966, *Torre & Correia* 14223 (COI, K, LISC, LMU, WAG). MS: Dondo Dist., 40 km from Beira on Chimoio (Vila Pery) road, fl. 30.xi.1965, *Torre & Correia* 13358 (COI, LD, LISC, LMA, LMU, MO, PRE, SRGH); Dondo Dist., Manga, fl. xii.1923, *Honey* 788 (K).

Not known elsewhere. Grasslands, sometimes seasonally flooded, sometimes with shrubs, on sandy or black alluvial soils near the coast; 20–50 m.

Conservation notes: Apparently endemic to central coastal Mozambique; probably Vulnerable.

N.E. Brown's name was published on 31 July 1906, according to a handwritten note in the Kew Library copy. The part of Bull. Soc. Bot. France that includes Gagnepain's name was received at Kew on 4 August 1906 and the part states "publié en juillet 1906". Allowing for packing and transit times between Paris and London, it must be reasonably certain that Gagnepain's name has priority, although the Paris library copy does not bear a date of receipt. I am grateful to Peter Phillipson for this information.

4. **Siphonochilus aethiopicus** (Schweinf.) B.L. Burtt in Notes Roy. Bot. Gard. Edinb. **40**: 372 (1982). —Lock in F.T.E.A., Zingiberaceae: 20 (1985). Types: Sudan, Fassogli, *Cienkowski* s.n. (B† syntype) and Ethiopia, Wochni (?Wahni), *Steudner* 496 (B† syntype, K).

   Cienkowskia aethiopica Schweinf., Beitr. Fl. Aeth.: t.1 (1867). —Solms in Schweinfurth, Beitr. Fl. Aeth.: 197 (1867).

   Kaempferia aethiopica (Schweinf.) Ridl. in J. Bot. **25**: 131 (1887). —Baker in F.T.A. **7**: 294 (1898). —Schumann in Engler, Pflanzenr. IV.46: 69, fig.10 (1904). —Troupin in Bull. Jard. Bot. État **25**: 267 (1955).

   Kaempferia ethelae J.M. Wood in Gard. Chron. **1898**(2981): 94 (1898). Type: Mozambique, Manica (Massi Kessi), Beningfield, cultivated at Natal Bot. Garden.

   Kaempferia dewevrei De Wild. & T. Durand in Bull. Soc. Bot. Belg. **38**: 142 (1899). Type: Congo, near Marioe Mt, ix.1896, *Dewevre* 1021a (BR holotype).

   Kaempferia evae Briq. in Ann. Conserv. Jard. Bot. Genève **6**: 3 (1902). Type: Zambia, ?Sefula dambo ("Pays des Ba-Rotsi: env. de Sefula"), *de Prosch* 12 (G holotype).

   Kaempferia zambesiaca Gagnep. in Bull. Soc. Bot. Fr. **53**: 355 (1906). Type: Mozambique, Campo, mouth of Zambezi, 22.xii.1904, *Le Testu* (BM holotype, EA, BR).

   Siphonochilus natalensis J.M. Wood & Franks in Medley Wood, Natal Plants **6**(3): t.560, 561 (1911); in Kew Bull. **1911**: 274 (1911).

   Siphonochilus evae (Briq.) B.L. Burtt in Notes Roy. Bot. Gard. Edinb. **40**: 372 (1982).

Perennial herb from a short ovoid rhizome 3–5 cm long. Roots with fusiform tubers 3–10 cm long. Leaves undeveloped or developing at anthesis; lamina narrowly elliptic, 17–36 × 2–4.5 cm, apex acuminate, base narrowly cuneate, quite glabrous; false petiole absent; ligule scarious, 6–35 cm long; sheaths furrowed, glabrous, forming a well-developed pseudostem to 70 cm tall. Inflorescence arising at base of leafy shoot, 4–12-flowered, with very short peduncle up to 2 cm long, mostly below ground; bracts oblong, obtuse, c.2.5 × 1 cm; floral bracts oblong, obtuse, 3–4 × 1 cm. Calyx tubular, shortly 3-lobed at apex, 2.7–3.5(5) cm long, glabrous. Corolla-tube 2.5–5.5 cm long, 1–2 mm wide; petals elliptic, 2.8–5.5 × 0.6–0.8 cm, whitish and translucent. Labellum mauve to purple, 3-lobed, 5–11.5 cm long, central lobe rounded and usually deeply emarginate with a central deep yellow mark at the base; lateral lobes broadly triangular, obtuse. Stamen obovate to oblong, rounded, sometimes laciniate or retuse at apex, up to 5 cm long; anthers 8–9 cm long. Ovary obovoid, 5–7 mm long, glabrous. Fruit ± subterranean, subglobose, weakly 3-lobed, c.1.5 cm wide, Seeds c.6 × 2 mm, pale brown, shiny.

Zambia. B: Kaoma Dist., 5 km W of Mankoya, 5.xi.1959, *Drummond & Cookson* 6231 (SRGH). N: Kasama, 10.xii.1961, *Astle* 1094 (SRGH). W: Kalulushi Dist., Chati Forest Reserve, 15.xii.1954, *Fanshawe* 9059 (SRGH). C: Mpika Dist., Mfuwe, Lubi R., 16.xii.1966, *Astle* 5002 (SRGH). E: Chipata Dist., Chikowa Mission, Luangwa Valley, 29.xi.1966, *Mutimushi* 1608 (K, SRGH). S: Choma Dist., 50 km N of Choma, fl. 11.xi.1954, *Robinson* 948 (K). **Zimbabwe**. N: Gurue Dist., W of Kanyemba, 28.i.1966, *Müller* 229 (K, LISC, SRGH). E: Chimanimani Dist., Charleswood Estate, 27.xi.1955, *Drummond* 5037 (BR, K, SRGH). **Malawi**. N: Mzimba Dist., Mzuzu, Marymount, 28.xi.1972, *Pawek* 6028 (SRGH). C: Dowa Dist., Mvera Mission (Uvera), 24.v.1901, *Kenyon* 25 (K). S: Zomba Dist., Malosa Mt, above Malosa Sec. School, 6.xi.1977, *Brummitt & Dudley* 15045 (K, SRGH). **Mozambique**. N: Mandimba Dist., Mandimba, fl. 6.xi.1941, *Hornby* 3455 (K). Z: Mopeia Dist., c.85 km from Morrumbala on road to Quelimane, 8.xii.1971, *Pope & Müller* 553 (LISC, SRGH). T: Angónia Dist., near Ulónguè, 1.xii.1980, *Macuácua* 1358 (K). MS: Mossurize Dist., Mt Espungabera (Spungabera), 21.xi.1960, *Leach & Chase* 10500 (K, SRGH).

Widespread in the savanna regions of tropical Africa from Senegal to Ethiopia and south to South Africa (KwaZulu-Natal). In deciduous woodland, bushland and wooded grassland; 350–1850 m.

Conservation notes: A very widespread species; not threatened. However, it is much sought-after as a medicinal plant in South Africa, where it is now very rare or extinct in the wild. If this use spreads, or an export trade develops, it could become Vulnerable, at least locally.

The species is variable and is treated here in a broad sense. Field studies, combined with cultivation to see the fully-developed vegetative parts, may reveal that there is more than one taxon in the region. Care is needed to avoid confusing white forms with *S. rhodesicus*; the key provides distinguishing features.

Kaempferia aethiopica Benth. var. *angustifolia* Ridley (type: *Welwitsch* 683 from Angola (BM, photo BR)) belongs to a related, undescribed species. Although the type lacks flowers, the lack of an elongated pseudostem excludes *K. aethiopica*. Two other specimens from Angola may well be the same: *Lynes* 329e, Missão de Luz, 27.xii.1933 (BR) and *Machado* ANG.XII.54-50, Cameia, xii.1954 (LISC).

5. **Siphonochilus rhodesicus** (T.C.E. Fr.) Lock in Kew Bull. **39**: 841 (1984); in F.T.E.A., Zingiberaceae: 20 (1985). Type: NE Zambia, Mukanshi R., *R.E. Fries* 1146 (UPS holotype).

   Kaempferia rhodesica T.C.E. Fr. in R.E. Fries, Wiss. Ergebn. Schwed. Rhod.-Kongo-Exped.: 236 (1916).

Perennial herb from a short subspherical rhizome. Roots not tuberous in any specimens seen. Leaves (mature ones not seen) narrowly elliptic to narrowly obovate, apex acute, aristate, base cuneate; ligule brown, membranous, to 5 mm or more long; sheaths furrowed, probably forming a pseudostem. Flowers appearing before the leaves, c.6 in each short congested inflorescence; floral bracts scarious, glabrous, chestnut-brown, narrowly ovate, to 4.5 × 1 cm. Calyx tubular, 3–7 cm long, with 3 obtuse apical teeth 3–5 mm long, split at one side to about a quarter its length. Corolla tube 3.5–13 cm long; petals narrowly oblong to narrowly elliptic, abruptly acuminate, 4.4–6 × 0.7–1 cm, usually ± ²/₃ the length of dried labellum. Labellum white with 2 central yellow marks, 6.5–9 × 7.5–9.5 cm overall, 3-lobed, the lateral lobes rhomboid, obtuse, 3.5–5 × 2.3–4 cm, central lobe broadly triangular, deeply emarginate at apex, 4–6 × 5.5–8.5 cm. Stamen obovate, obtuse, c.3.5 × 0.6 cm; anther basal, c.1 cm long. Ovary oblong, c.8 mm long. Fruit and seeds unknown.

Zambia. N: Mbala Dist., 8 km S of Mbala (Abercorn) on Kasama road, 19.x.1947, *Greenway & Brenan* 8234 (K). W: Kitwe, 14.x.1970, *Fanshawe* 10938 (K). **Malawi**. N: Nkhata Bay Dist., North Vipya, road from Njakwa to Uzumara Forest, 29.xi.1970, *Pawek* 4054 (K, MAL).

Also in Tanzania and Congo. In deciduous woodland, usually with *Brachystegia*, often on hillsides; 1350–1950 m.

Conservation notes: Common and with a fairly wide distribution; not threatened.

This is a poorly known species. The mature leaf form will remain unknown until specimens are cultivated. The relative lengths of the corolla lobes and the labellum should distinguish it from white forms of *S. aethiopicus*. According to Plowes & Drummond (Wild Fl. Rhodesia, 1976) it flowers at night, as does *S. longitubus*.

6. **Siphonochilus parvus** Lock in Kew Bull. **46**: 269 (1991). Type: Tanzania, Mufindi, Irundi Hill, 23.xi.1985, *Congdon* 46 (K holotype).

   Siphonochilus sp. D of Lock in F.T.E.A., Zingiberaceae: 21 (1985).

Perennial herb from a short upright rhizome. Roots bearing fusiform tubers to 8 cm long and 0.5 cm wide. Leaves developing after flowering; lamina narrowly elliptic, 24–30 × 2–2.5 cm, apex acuminate, base narrowly cuneate, glabrous on both surfaces; false petiole absent; ligule brownish, membranous, 6–10 mm long, asymmetric, acute; sheaths furrowed, glabrous, forming a well-developed pseudostem to 10 cm long. Inflorescence arising at base of leafy shoot before the leaves appear, 9–14-flowered; peduncle very short; bracts oblong, to 1.5–0.5 cm; floral bracts oblong, 2.5–4.5 × 0.5 cm; pedicels 1.5–2 cm long. Calyx tubular, 2.1–2.5 cm long, shortly 3-lobed at apex, lobes usually with dorsal subapical subulate appendage. Corolla tube 2.5–3 cm long; petals narrowly elliptic, 2–2.5 × 0.4–0.6 cm. Labellum suborbicular in outline, 3-lobed, 2.2–3.2 cm long; lateral lobes ± triangular, acute, central lobe deeply emarginate, ± rounded. Stamen c.1.7 × 0.2 cm, narrowly rectangular, apex rounded; anther 4–5 mm long. Ovary ovoid, 3–4 mm long. Young fruit spherical, 1 cm wide. Seeds (slightly immature) smooth, obovoid to ellipsoid, greyish-brown.

Zambia. N: Mbala Dist., Isanya Coffee Farm, fl. 29.x.2004, *Bingham & Hemmings* 12804 (K). W: Mwinilunga Dist., S of Matonchi Farm, fl. 12.x.1937, *Milne-Redhead* 2719 (K, BR); Ndola Dist., Nkumbo (Chibuli) Hill, 13°24'00"S 28°14'30"E, 11.xi.1998, *Leteinturier, Malaisse & Matera* 422 (BR). **Malawi**. N: Nyika Nat. Park, c.5 km N of Thazima, fl. 24.x.1986, *E.A.S. & I.F. la Croix* 867 (K).

Also in Tanzania. Rocky grassland in deciduous woodland; 1300–1900 m.

Conservation notes: Apparently an uncommon species; there are few definite collections but this may be because the flowering season is short and the flowers relatively inconspicuous. Probably either Data Deficient or Vulnerable.

7. **Siphonochilus puncticulatus** (Gagnep.) Lock, comb. nov. Type: W Zambia?, "Haut-Zambeze, à Lefula", xi, *Kiener* s.n. (P holotype).

Kaempferia puncticulata Gagnep. in Bull. Soc. Bot. Fr. **53**: 353 (1906). —Troupin in Bull. Jard. Bot. État. **25**: 267 (1955).

Kaempferia homblei De Wild. in Repert. Spec. Nov. Regni Veg. **13**: 195 (1914). Types: Congo, Plateau de Biano, xii.1912, *Homblé* 907 (BR syntype) & Congo, around Katentania, xi.1912, *Homblé* 851 (BR syntype).

Translation of original Latin description: "Rhizome creeping, brown-scaly, bearing tuberous napiform blackish roots (thickened part 2 × 0.8 cm) constricted on both sides [at both ends?]. Leafy shoot medium (20 cm tall), at the base bearing small short sheaths, the upper ones larger but lacking laminas. Leaf sheath open, glabrous, folded; ligule narrow, rounded at apex; lamina narrow, 15–18 × 2.5–3 cm, lanceolate, elongate, acute at apex, narrowed acuminately at base into a petiole above the sheath, smooth, glabrous, pale green. Flowering stems few or solitary, erect from base of leafy shoot and exceeded by it, bearing bracts that are short near base and longer above. Flowers solitary (in our material) or few, very large, subtended at base by thinly papery bracts. Ovary glabrous, hardly 1 cm long. Calyx 3.5 cm long, tube 1.8 cm long, elongate-turbinate, thinly papery, brown-punctulate outside, rugose-plicate, 3-lobed, lobes ovate, 1.9 cm long, separated by sinuses 5–7 mm deep. Corolla brown-punctulate outside, the tube markedly dilated above; petals subequal, 2.5–3 cm long, ovate or oblong, apex acute or acuminate. Labellum very broad, 4 cm longer than petals, trilobed, lobes all irregularly erose-crenulate, a beautiful lilac-pink. Anthers 5–6 mm long, oblong-linear, exceeding calyx; anther crest oblong, 6 × 0.6–0.7 cm, trilobed at apex, the middle and largest lobe rounded; staminodes small, obtuse".

Zambia. W: Kabompo Dist., 55 km W of Kabompo, 26.xii.1969, *Simon & Williamson* 2040 (SRGH) may be this species. Photographs (now at K) by Alex Paul at Zambia B: Kalabo Dist., Liuwa Plain, 14°30'S 22°30'E appear to show this species, but there are no accompanying specimens.

Also in Congo and Angola. Apparently confined to Kalahari sands; 1200–1400 m.

Conservation notes: Poorly known from few specimens; Data Deficient.

The type sheet at P ("Haut-Zambeze, Mlle Kiener, recu de M. se Seynes le 18 Juin 1896, Lefula [could be read as Sefula], premieres ondees de Novembre") has been annotated by Troupin, x.1955: "K. homblei De Wild. ...est identique et doit etre considere comme synonyme. Voir Bull. Jard. Bot. État. 25(4): Dec. 1955)".

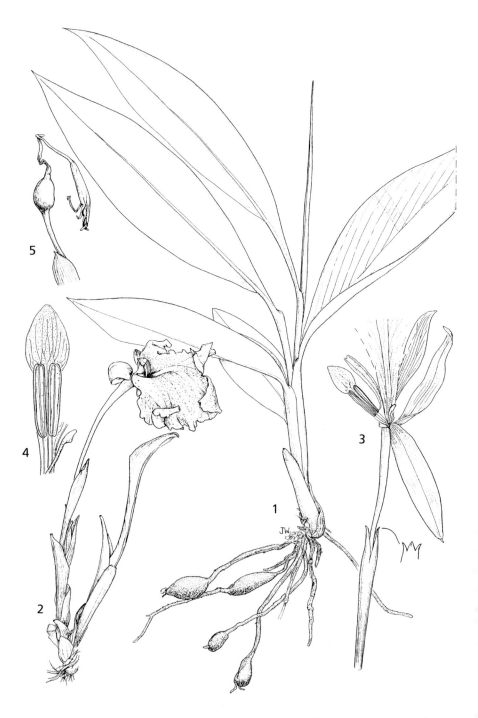

Fig. 13.4.**23**. SIPHONOCHILUS LONGITUBUS. 1, habit (× ¹/₂); 2, flower habit (× ²/₃); 3, details of flower (×1); 4, stamen (× 2); 5, developing fruit (× 1). 1–2 from *Robinson* 383, 3–5 from *Drummond & Cookson* 6759. Drawn by Juliet Williamson.

This is an exceptionally unsatisfactory inclusion in the Flora. It is clear from Alex Paul's photographs that there is a distinct species in the Liuwa Plains Nat. Park; the photos show pink-mauve precocious flowers that appear to be much pinker than those of *S. aethiopicus* with distinctive frilly margins to the labellum, and developing leaves that show no sign of forming a pseudostem. The species that most closely resembles this is Gagnepain's *Kaempferia puncticulata*. Some specimens that have been named as this, as well as the type of *Kaempferia evae*, show a pseudostem clearly beginning to form and must be wrongly named; they are probably forms of *S. aethiopicus*.

The few records of plants that probably belong to this species are all from Kalahari sand areas. Dried material shows numerous small brown dots, perhaps oil glands, in the tissues of the labellum, corolla and calyx. Good material of this species, particularly grown on to show the mature leaf form, is much needed.

8. **Siphonochilus longitubus** Lock, sp. nov. Type: Zambia, Choma Dist., Mapanza Mission, fl. 25.xi.1953, *Robinson* 393 (K holotype)[3]. FIGURE 13.4.**23**.

Perennial herb. Roots bearing ellipsoid tubers 1.5–2.5 × 0.6–1 cm, 5–8 cm from stem base. Young leaves often (but not always) present at flowering. Ligule c.1 mm long, rounded. Lamina narrowly ovate to elliptic, 11.5–20 × 2.5–4.5 cm (perhaps larger when fully developed), glabrous, base narrowly cuneate with short false petiole to 1 cm long, apex acuminate. Flowers borne on a very short condensed inflorescence; peduncle not visible. Calyx tubular, 2.8–4.5 cm long, split down one side for 9–13 mm at the apex. Corolla tube 6–9 cm long and c.1 mm wide; lobes narrowly oblong, 24–39 × 4–7 mm, apiculate at apex, often curled back after anthesis, reddish brown with numerous darker dots near apex and near the margins towards apex. Labellum white, suborbicular in outline, deeply 3-lobed, central lobe itself bifid, c.6 × 4.5 cm. Stamen oblong, obtuse to rounded at apex, 2.1–2.4 × 0.5–0.7 cm; anther cells 9–10 × 1 mm, broadening slightly upwards. Stigma trumpet-shaped, 2 mm broad at apex. Fruit and seeds not known.

Zambia. W: Ndola, fl. 13.xi.1953, *Fanshawe* 499 (BR, K, NDO). C: Kafue W, Kasusu R., fl. 30.xi.1997, *Bingham & Luwiika* 11609 (K). S: Choma Dist., Mapanza Mission, fl. 25.xi.1953, *Robinson* 393 (K).

Not known elsewhere. Zambesian woodland, often on termitaria; 980–1100 m.

Conservation notes: Endemic to south-central Zambia, but not threatened. Its night-flowering habit means that it is probably undercollected.

Although *Siphonochilus longitubus* has white flowers like those of *S. rhodesicus*, it appears to be morphologically closer to *S. brachystemon*, a blue-flowered species of dry forests of which there is only one collection in the Flora area. The form of the sterile portion of the stamen and the subspherical tubers are similar in the two species. *Siphonochilus rhodesicus*, on the other hand, is closest to *S. aethiopicus*, differing only in the proportions of the flowers. The distributions of *S. longitubus* and *S. rhodesicus* do not appear to overlap.

The white flowers are recorded (*Drummond & Cookson* 6759) as opening in the evening and in most collections the labellum has shrivelled. The flowers of *Best* 46 are described as "predominantly white with a slight pinkish tinge and a yellow centre [the labellum]. 'Flower stem' [the corolla tube] pinkish with greenish sheath [calyx]. 3 sepals [petals] pinkish. Has a distinct and pleasant odour". The mature leaf form will remain unknown until specimens are collected and grown on to the mature stage.

---

[3] *Siphonochilus longitubus* Lock, sp. nov., *S. brachystemoni* (K. Schum.) B.L. Burtt forma partis sterilis staminis, tubo corollae longissimo, radicibus tuberis subsphaericis munitis similis sed floribus albis non caeruleis, foliis hysteranthis non coaetaneis, petiolo breviore c. 1 cm non 3–8 cm longo differt. Typus: Zambia, Mapanza Mission, *Robinson* 393 (K).

Other material seen: *Best* 46 (K), *Drummond & Cookson* 6200 (K, LISC), *Drummond & Cookson* 6759 (K, LISC), *Fanshawe* 1728 (K, BR), *Fanshawe* 9863 (K), *Fanshawe* 11142 (K), *King* 245 (K), *Mrs Macauley* 955 (K), *Shepherd* 86B (K), *Trapnell* CAS 511 (K), *West* 3454 (K).

9. **Siphonochilus brachystemon** (K. Schum.) B.L. Burtt in Notes Roy. Bot. Gard. Edinb. **40**: 372 (1982). Types: Tanzania, Tanga Dist., Mkulumuzi (Mukulumussi), *Volkens* 201 & Lushoto Dist., Sigi, *Holst* 3100 (B† syntypes).

> *Kaempferia brachystemon* K. Schum. in Engler, Pflanzenw. Ost-Afr. **C**: 149 (1895); in Engler, Pflanzenr. IV. **46**: 71, fig.10D (1904). —Baker in F.T.A. **7**: 296 (1898).
>
> *Cienkowskiella brachystemon* (K. Schum.) Y.K. Kam in Notes Roy. Bot. Gard. Edinb. **38**: 10 (1980).
>
> *Kaempferia macrosiphon* Baker in F.T.A. **7**: 296 (1898). Type: Kenya, 'Nyika country', near Mombasa, i.1879, *Wakefield* (K holotype).

Perennial herb from a short rhizome. Roots with subspherical tubers c.5 mm diameter when dry. Leaves clustered, usually fully developed at flowering time; lamina elliptic, glabrous, somewhat asymmetric, acuminate, base cuneate, 16–37 × 6.5–12 cm; false petiole 3–8 cm long; ligule short, membranous; sheaths 6–10 cm long; pseudostem very short or absent. Inflorescence arising at base of leafy shoots, 5–12-flowered; peduncle 0.5–4 cm long; bracts scarious, delicate, narrowly elliptic, to 3.5 cm long; floral bracts smaller, to 1 cm long but delicate and usually poorly preserved; pedicels 0.5–1(5) cm long. Calyx tubular, 2.5–6.5 cm long, with 2 long and 1 short tooth at apex. Corolla tube 4.5–10 cm long, 1–2 mm wide; petals elliptic, acute, 2–4 × 0.5–0.8 cm, blue. Labellum blue, suborbicular in outline, deeply 3-lobed, with the central lobe itself bifid, 3–5 × 4.5–8 cm. Stamen c.1.5 cm long, obovate, obtuse; anther 5 mm long. Style filiform, stigma flattened, with a dorsal ridge, suborbicular. Ovary narrowly ovoid. Fruit and seeds unknown.

Mozambique. N: Montepuez Dist., 3 km from Montepuez towards Nantulo, 27.xii.1963, *Torre & Paiva* 9720 (BR, COI, LISC, LMA, LMU, MO, SRGH).

Also in Sudan, Uganda, Kenya, Tanzania, Burundi and Congo. Herb layer of gallery forest and drier evergreen forests; 530 m.

Conservation notes: Within the Flora area known only from a single collection from gallery forest, a threatened habitat; probably Vulnerable. Globally, it is a species of drier forests but relatively widespread; perhaps Near Threatened.

3. **AFRAMOMUM** K. Schum.

Aframomum K. Schum. in Engler, Pflanzenr. IV. **46**: 201 (1904). —Burtt & Smith in Notes Roy. Bot. Gard. Edinb. **31**: 177–228 (1972). —Lock & Poulsen in Kew Bull. **52**: 601–616 (1997).

Perennial herbs with erect leafy stems from extensive rhizome systems. Leaves alternate, with long sheathing bases forming a pseudostem; sheaths furrowed or reticulate; ligule present; base of lamina contracted into a pseudopetiole; laminas of lowest leaves reduced, upper laminas ovate to obovate, usually glabrous above, variously pubescent beneath, or glabrous; lateral nerves easily visible beneath but sometimes obscured by hypodermal sclerenchyma above. Inflorescences terminating separate leafless shoots arising either at the base of leafy shoots or from the rhizomes, or from both; peduncle with 2 rows of bracts; flowers clustered at apex of peduncle, each subtended by 2 subopposite bracts; flowers large, strongly zygomorphic. Calyx spathaceous, splitting unilaterally at anthesis. Lateral petals narrow, often recurved, posterior petal larger, usually oblong and concave, forming a hood-like structure; basal part of petals and androecium fused into a tube. Androecium consisting of a single free stamen with subulate outgrowths at base, apex usually 3-lobed with 2 subterminal anther-thecae dehiscing by longitudinal slits, and a labellum that may originate from 1 or 2 staminodes. Labellum either broad and delicate or narrower, thicker and decurved. Ovary inferior, trilocular; style filiform, passing between the anther-thecae, ending in a trumpet-shaped stigma with a ciliate margin. Subulate or contorted nectaries ('stylodia') occur at style base. Fruit a large indehiscent berry with a tough fleshy wall, usually red when ripe, with an apical beak formed from the persistent calyx, style base, or a sterile prolongation of the ovary. Seeds hard, black or brown, surrounded by white translucent arils, cohering into 3 masses around the delicate axile placenta.

A genus of about 50 species, mainly growing in forest in the wetter parts of tropical Africa from Senegal and Ethiopia to Mozambique and Angola.

Close to some species of *Amomum* Roxb. from tropical Asia, but cladistic analyses have shown that they are distinct. Other studies (Harris *et al.*, Edinb. J. Bot. **57**: 377–395, 2000) have shown that diversification in the genus is recent with most of the speciation occurring in the last 1–2 million years. The genus is currently being revised by David Harris (E); I am grateful for his comments and assistance.

The flowers of both *Aframomum* and *Siphonochilus* are very delicate and open for a single day or part of a day. Collections that included carefully dissected and dried flowers are very useful.

Hypodermal sclerenchyma, mentioned in the generic description and in some of the species descriptions, consists of strands of thick-walled cells that lie just beneath the upper epidermis of the leaf. It is visible in dried material with a good lens and appears as a large number of minute discontinuous veins.

1. Inflorescence capitate, comprising more than 20 rather small flowers; labellum less than 3 cm long; nectaries forming a contorted mass at style base · · · · · · · 2
– Inflorescence of fewer large flowers; labellum more than 3 cm long; nectaries paired, subulate · 3
2. Leaves pubescent only on margins and midrib below; labellum whitish with dark red centre · **1.** *zambesiacum*
– Leaves pubescent over whole lower surface of lamina; labellum orange-yellow · **2.** *polyanthum*
3. Labellum relatively firm, decurved, red, orange or yellow; anthers dehiscing along their whole length · **3.** *angustifolium*
– Labellum very delicate, erect or spreading, pink, mauve or white; anthers dehiscing for about half their length · 4
4. Inflorescence arising at base of a leafy shoot (which may be represented only by a burned stump); ligule bifid, pubescent with a mixture of short hairs and reddish glands; midrib of lamina pubescent beneath; upper surface of lamina with hypodermal sclerenchyma · **4.** *alboviolaceum*
– Inflorescence arising from rhizomes between the leafy shoots; ligule rounded, sparsely pubescent with short hairs only; whole undersurface of lamina sparsely pubescent; upper surface of lamina without hypodermal sclerenchyma · **5.** *albiflorum*

1. **Aframomum zambesiacum** (Baker) K. Schum. in Engler, Pflanzenr. IV. **46**: 206 (1904). —Lock in Bull. Jard. Bot. Belg. **48**: 130 (1978); in F.T.E.A., Zingiberaceae: 25 (1985). Type: Malawi, Zomba Mt, 1.x.1859, *Kirk* s.n. (K holotype, P).
Amomum zambesiacum Baker in F.T.A. **7**: 309 (1898).

Subsp. **zambesiacum**. —Lock in Bull. Jard. Bot. Belg. **48**: 130 (1978).

Perennial herb. Rhizome 8–12 mm in diameter, short and much-branched so plant grows in dense clumps. Leafy stems to 2 m tall. Leaves broadly elliptic, 20–45 × 5–14 cm, caudate-acuminate, base cuneate, glabrous except for margins and midrib which have dense appressed ascending hairs beneath; veins widely spaced, 5–7 per 5 mm above, 13–19 per 5 mm beneath; ligule coriaceous, round, to 5 mm long; leaf sheaths furrowed, glabrous. Inflorescence 25–50-flowered, arising from base of leafy shoots; peduncle 10–40 cm long; bracts oblong, sheathing, papery to coriaceous, 3.5–5 × 2–2.5 cm. Calyx spathaceous, 1.5–2 cm long, split for c.6 mm on one side, sparsely appressed-pubescent outside, densely so near apex, glabrous inside. Petals whitish, the posterior ovate, c.1.8 × 1 cm, the laterals oblong, c.1.8 × 0.7 cm, creamy white. Labellum 3.5–4 cm long, the free portion spathulate, c.2 × 1.5 cm, creamy white with a large dull crimson patch at the base. Free filament 10–12 mm long, ligulate, apex retuse; anther subterminal, c.9 mm long, almost completely dehiscent. Style filiform, with scattered straight appressed hairs; stigma trumpet-shaped, densely hairy. Stylodia forming a contorted mass around style base. Fruits ovoid, c.7 × 4–5 cm, thick-walled, with prominent longitudinal ridges, red when fresh. Seeds irregularly ellipsoid, 5 × 4 mm, dark brown, shiny, irregularly colliculate.

Malawi. N: Mzimba Dist., Mzuzu, 26.x.1976, *Pawek* 11916 (K, MO, SRGH). S: Thyolo Dist., Thyolo (Cholo), 19.ix.1949, *Wiehe* N/236 (SRGH).

Also in Nigeria, Cameroon, Uganda, Kenya and Tanzania. Upland forest and secondary growth, often near paths or streams; 1500–2000 m.

Conservation notes: Widely distributed in upland tropical Africa although its montane forest habitat is vulnerable to clearance and cultivation; possibly Near Threatened.

A separate subspecies with densely puberulous peduncle bracts, subsp. *puberulum* Lock, occurs in Ethiopia.

2. **Aframomum polyanthum** (K. Schum.) K. Schum. in Engler, Pflanzenr. IV. **46**: 207 (1904). —Hallé in Adansonia, ser.2, **7**: 73 (1967). —Lock in Bull. Jard. Bot. Belg. **48**: 130 (1978); in Kew Bull. **35**: 302 (1980). Type: Sudan/Congo border, Diamvonu, Niamniam, *Schweinfurth* 3262 (K lectotype, designated by Lock 1978).

  Amomum polyanthum K. Schum. in Bot. Jahrb. Syst. **15**: 411 (1893).

Perennial herb. Rhizome extensive, c.30 mm diameter. Leafy shoots 3–4 m tall. Leaves up to 55 × 19 cm, broadly ovate-elliptic, abruptly caudate-acuminate, base cuneate, densely pubescent on midrib and margins beneath, rather obscurely appressed-pubescent between nerves beneath; nerves above 0.5–1.5 mm apart; sheaths longitudinally ridged, pubescent. Inflorescence 10–35 cm tall, capitate, 10–26-flowered, arising from base of leafy shoot; bracts distant, glabrous; flowers surrounded by overlapping bracts. Calyx 3–4 cm long, puberulous, red, split irregularly into 3 lobes towards apex. Corolla 1.8–2 cm longer than calyx, red, the posterior lobe brighter. Labellum 2.5 cm wide, reflexed with a frilly margin, orange-yellow. Fertile stamen with anther loculi 9.5 mm long; anther crest with a small rounded to emarginate central lobe and linear decurved lateral lobes. Fruits several together at apex of peduncle, longitudinally c.9-ridged, spherical to ovoid, c.3.5 cm wide, red. Seeds c.9 × 3 mm, narrowly ovoid, acuminate, shiny, finely striate, irregularly covered in small rounded protuberances.

Zambia. B: Kabompo Dist., Kabompo, fl. 22.ix.1964, *Fanshawe* 8920 (BR, K, LISC). N: Kawambwa Dist., near Kawambwa Boma, fl. 30.x.1952, *White* 3545 (K).

Also in Cameroon, Gabon, Congo, Sudan and NW Uganda. Swamp and riverine forest; 1200–1400 m.

Conservation notes: South of the Equator known only from these two collections; best treated as Data Deficient.

The description is taken largely from material from north of the Equator. White describes the flowers as orange, while Fanshawe says they are yellow. In Uganda the flowers are red and orange.

3. **Aframomum angustifolium** (Sonnerat) K. Schum. in Engler, Pflanzenr. IV. **46**: 218, fig.27 (1904). —Robyns & Tourney in Fl. Parc Nat. Albert **3**: 406 (1955). —Burtt & Smith in Notes Roy. Bot. Gard. Edinb. **31**: 188 (1972). —Lock in Kew

Bull. **35**: 302 (1980); in Kew Bull. **39**: 840 (1984); in F.T.E.A., Zingiberaceae: 27 (1985). Type: Madagascar, *Sonnerat* s.n. (BM).

Amomum angustifolium Sonn., Voy. Indes Orient. **2**: 276, t.136 (1782). —Baker in F.T.A. **7**: 308 (1898) in part.

Perennial herb. Rhizome extensive, 5–15 mm diameter; scales coriaceous, red-brown, to 7 cm long. Leafy shoots 1.5–4 m tall. Leaves oblong-elliptic, 30–55 × 6.5–12.5 cm, abruptly acuminate, base cuneate, slightly asymmetric, ± pubescent on midrib beneath; nerves beneath (20)24–29 per 5 mm, obscured above by hypodermal sclerenchyma; ligule coriaceous, rounded, 4–10 mm long; sheaths furrowed with some pitting within the furrows, glabrous. Inflorescence arising at base of leafy shoots, simple or branched, 4–10-flowered; peduncles (10)30–70 cm long, erect, basal bracts broadly ovate, 2 × 1.5 cm; uppermost bracts broadly ovate, 5 × 4 cm, concave, glabrous, emarginate and mucronate, coriaceous with scarious margins. Calyx spathaceous, 3.5–4.5 cm long, split 1.5–3 cm down one side. Petals red, the posterior broadly ellipsoid, concave, 3.2–4.4 × 1.6–2.2 cm, laterals narrowly ovate, 2–3.5 × 0.5–0.7 cm. Labellum c.7.5 cm long, free portion 3.8–4.5 × 2.2–2.5 cm, spathulate, decurved, yellow or pale orange. Free filament ligulate, 2.5–2.9 cm long; anther 10–14 mm long, completely dehiscent; apex 3-lobed, the terminal lobe triangular, 2–3.5 mm long, lateral lobes subulate, 4–8 mm long. Style glabrous or slightly pilose, stigma trumpet-shaped, externally pubescent. Fruits ovoid to subglobose, 7–9.5 × 3.2–3.7 cm, including persistent calyx. Seeds ovoid, 4–5 × 2–3 mm, smooth.

Zambia. N: Mpika Dist., Mpika, 27.i.1955, *Fanshawe* 1869 (K). **Zimbabwe**. E: Chimanimani Dist., Glencoe Forest Reserve, 22.xi.1955, *Drummond* 4956 (SRGH). **Malawi**. N: Nkata Bay Dist., 8 km E of Mzuzu at Roseveare's, 1300 m, 1.xi.1969, *Pawek* 2936 (K). C: Nkhotakota Dist., Nchisi Forest, 12.xii.1962, *Adlard* 522 (K). S: Blantyre Dist., Bangwe Hill, 4 km E of Limbe, 23.xi.1977, *Brummitt, Seyani & Banda* 15154 (K, MAL, SRGH). **Mozambique**. N: Mueda Dist., 30 km from Mueda to Diaca, 15.iv.1964, *Torre & Paiva* 11988 (COI, LISC. LMA, LMU, PRE, WAG). Z: Maganga da Costa Dist., 30 km from Vila da Maganga on road to Bajone, 120 m, 23.xi.1967, *Torre & Correia* 16228 (BR, LISC, LMU). MS: Sussundenga Dist., Chimanimani Mts, path between Skeleton Pass and Namadima, 30.xii.1959, *Goodier & Phipps* 351 (K). M: Marracuene Dist., Bobole Reserve, c.15 km past Marracuene on road to Xai-Xai, 30.x.1981, *Jansen & Macuácua* 7792 (K).

Widespread in tropical Africa from Ivory Coast to Sudan and south to Mozambique; also in Madagascar and the Mascarenes. Forest undergrowth and forest margins, often near water; 120–1400 m.

Conservation notes: A widespread species, although may be local in occurrence; not threatened.

The fruits were formerly traded as cardamoms and the range may have been artificially extended.

4. **Aframomum alboviolaceum** (Ridl.) K. Schum. in Engler, Pflanzenr. IV. **46**: 207 (1904). —Lock in Bull. Jard. Bot. Belg. **49**: 179 (1979); in F.T.E.A., Zingiberaceae: 34 (1985). Type: Angola, Pungo Andongo, R. Cuanza, xii.1856, *Welwitsch* 6453 (BM holotype). FIGURE 13.4.**24**.

Amomum alboviolaceum Ridl. in J. Bot. **25**: 130 (1887). —Baker in F.T.A. **7**: 304 (1898).

Aframomum biauriculatum K. Schum. in Engler, Pflanzenr. IV. **46**: 207 (1904). —Fanshawe in Kirkia **8**: 151 (1972). Types: Angola, Malange, i.1883, *Buchner* 7 & *Marques* 66 (B† syntypes).

Aframomum candidum Gagnep. in Bull. Soc. Bot. France **53**: 351 (1906). Types: Mozambique, 10.i.1904, *Le Testu* 606, 675 & 707 (BM, P syntypes).

Perennial herb. Rhizome extensive, c.8 mm diameter, buried at least 5 cm deep; scales delicate, soon disintegrating. Leafy shoots to 1.5 m tall. Leaves narrowly ovate, up to 32 × 9 cm, acuminate, base cuneate, midrib pubescent beneath, otherwise glabrous except for a row of minute tooth-like marginal hairs from raised bases; nerves 23–33 per 5 mm beneath, usually

Fig. 13.4.**24**. AFRAMOMUM ALBOVIOLACEUM. 1, habit and inflorescence (× $^1/_2$); 2, leaf base and short ligule (× $1^1/_2$); 3, leaf base and long ligule (× $1^1/_2$); 4, bract, side view (× 1); 5, flower, side/back view and calyx (× $^2/_3$); 6, stamen and style (× $1^1/_2$); 7, anthers and style (× 3). 1–2 from *Fanshawe* 2933, 3–7 from *Leach & Williamson* 13518. Drawn by Juliet Williamson.

obscured by hypodermal sclerenchyma above. Ligule coriaceous, bifid, the halves unequal, rounded to acute, the longer half 3–10(16) mm long, with a mixture of simple appressed hairs and reddish glands that collapse and become blackish after long storage; leaf sheaths often furrowed, often reticulate towards apex, with an indumentum like the ligule. Inflorescence 2–5-flowered, arising at the base of an old leafy shoot, often represented only by a charred base; peduncle bracts broadly ovate, up to 3 × 2.5 cm, usually pubescent, at least near apex, with mixed indumentum like the ligule and leaf sheath. Calyx spathaceous, 4–5 cm long, split for 1.5 cm down one side. Petals pale mauve to almost white, the posterior oblong, concave, rounded, 4.5–6 × 2–3.5 cm; laterals narrowly ovate, 4–6 × 0.7–1.4 cm. Labellum c.8 cm long, free portion suborbicular, c.6 cm wide, pale mauve to almost white, with a yellow patch at the base. Free filament c.2 cm long, ligulate; anther 9–12 mm long, lower half dehiscent; apex 3-lobed, apical lobe triangular with bifid apex, 2.5 × 5 mm, lateral lobes subulate, to 9 mm long. Style appressed-pubescent, stigma trumpet-shaped, fringed with hairs. Fruit ovoid, up to 10 × 7 cm, smooth with short beak composed mainly of base of calyx tube. Seeds ellipsoid, 4–6 × 3–4 mm, dark brown, shiny.

Zambia. B: Zambezi (Balovale), vii.1933 *Trapnell* 1255 (BR, K). N: Mbala Dist., 5 km from Kalambo Falls on road to Mbala, fl. 12.i.1975, *Brummitt & Polhill* 13733 (K). W: Ndola, fl. 21.ix.1954, *Fanshawe* 1678 (BR, K, SRGH). C: Mumbwa Dist., Mumbwa, Kafue Basin Survey, 18.v.1963, *van Rensburg* 2173 (K). **Malawi**. N: Mzimba Dist., Mzuzu, Marymount, *Pawek* 3019 (K). **Mozambique**. N: Mueda Dist., 15 km from Mueda to Chomba, 800 m, 15.iv.1964, *Torre & Paiva* 12025 (LISC, LMU). Z: Namacurra Dist., Namacurra, 10 km on road from Maganja da Costa, 40 m, 26.i.1966, *Torre & Correia* 14112 (BR, COI, LISC, LMA). MS: Dondo Dist., Manga, 1923, *Honey* 787 (K, PRE).

Widespread in tropical Africa from Senegal and Ethiopia in the north to Mozambique in the south. In moister woodlands and wooded grasslands, often close to the forest-savanna boundary; 40–1400 m.

Conservation notes: Widespread and often abundant species; not threatened.

5. **Aframomum albiflorum** Lock in Kew Bull. **39**: 839 (1984). Type: Malawi, Mt Mulanje, Ruo Gorge, 26.i.1967, *Burtt & Hilliard* 4634 (E holotype).

Perennial herb. Rhizome extensive, 6–8 mm diameter; scales coriaceous, reddish brown, c.3.5 × 1.5 cm. Leafy shoots 2–3 m tall. Leaves narrowly elliptic, acuminate, somewhat asymmetrically rounded to cuneate at base, 25–33 × 4.5–8 cm, pubescent on midrib beneath; nerves 22–27 per 5 mm beneath, 20–23 per 5 mm above; ligule coriaceous, rounded, sparsely pubescent, 3–6 mm long. Inflorescence 2–3-flowered, borne on rhizomes between the leafy shoots; peduncles 3–5 cm long, bracts broadly ovate, retuse and mucronate at apex, sparsely appressed-pubescent, ciliolate at apical margin, apical bracts up to 3.5 × 3 mm. Calyx spathaceous, 4–5 cm long, split c.1.5 cm at one side. Corolla white, posterior petal oblong, concave, 5.5–6 × 2.5–2.8 cm, laterals narrowly triangular, c.5.5 × 0.5 cm. Labellum white with a basal yellow patch, 9–10 cm long, free portion suborbicular, c.6 cm wide. Free filament ligulate, c.3 cm long; anther subterminal, c.1.5 cm long, the basal 10 mm dehiscent; apex 3-lobed, terminal lobe oblong, rounded or retuse, 6–7 × 4 mm, the laterals narrowly triangular, ascending, c.6 mm long. Ovary glabrous; style glabrous, stigma trumpet-shaped, appressed-pubescent outside. Fruit and seeds not known.

Zimbabwe. E: Chipinge Dist., Chirinda Forest, 1.xii.1961, *Wild* 5536 (COI, K, M, MO, SRGH). **Malawi**. N: Nkhata Bay Dist., 8 km E of Mzuzu, Roseveare's, 1.xi.1969, *Pawek* 2938 (K, MO, SRGH). S: Mt Mulanje, Ruo Gorge, 26.i.1967, *Burtt & Hilliard* 4634 (E). **Mozambique**. N: Ribáuè Dist., Serra de Ribáuè, Mepáluè, 1250 m, 25.i.1964, *Torre & Paiva* 10216 (K, LISC, LMU). Z: Milanje Dist., Mt Chiperone, upper SE slopes, 1150 m, 27.xi.2006, *Timberlake* 4900 (K, LMA). MS: Mossurize Dist., Mt Espungabera, 21.xi.1960, *Leach & Chase* 10511 (BM, COI, LISC, M, MO, PRE, SRGH).

Also in Tanzania. Montane forest undergrowth and forest margins; 1000–1500 m.
Conservation notes: The species has a fairly extensive distribution but in a restricted montane forest habitat which is often under pressure. As species of *Aframomum* often flourish after disturbance, Near Threatened is probably appropriate.

CANNACEAE

by J.M. Lock & M.A. Diniz

Perennial herbs arising from horizontal thickened rhizomes. Leafy shoots erect. Leaves spirally arranged with open sheaths; petiole and ligule absent. Inflorescence terminal, pedunculate, many-flowered, bracteate, either a simple spike or a pseudo-spike, or a thyrse made up of 2-flowered subunits. Flowers bisexual, trimerous, asymmetric; perianth segments (tepals) 6, in 2 whorls. Outer whorl with 3 ovate segments, free, persistent; inner whorl with 3 narrowly oblong segments, sometimes unequal, longer than those of outer whorl, shortly united at base. Androecium petaloid with 0–3 staminodes united below to form a tube, usually with the anterior staminode (labellum), much larger than the others; stamen 1, petaloid, with 1 fertile theca. Ovary inferior, trilocular with axile placentation, surface papillose, spiny-fimbriate or verrucose; ovules numerous; style fleshy, petaloid, united to staminodes and petaloid part of the stamen. Fruit a thin-walled capsule, spiny or tuberculate, dehiscing irregularly loculicidally, sometimes tardily so, many-seeded. Seeds globose, smooth with hard endosperm.

A family with 1 genus of about 25 species, mostly native to tropical and subtropical America and a few in the Old World tropics. The status of the African plant is unclear but it is probably an early introduction. Various hybrids (*Canna × generalis* L. Bailey) and forms of *C. indica* are cultivated.

CANNA L.

Canna L., Sp. Pl.: 1 (1753). —Segeren & Maas in Acta Bot. Neerl. **20**: 663–680 (1971).

Description as for the family.

Canna indica L., Sp. Pl.: 1 (1753). —Hepper in F.W.T.A. ed.2, **3**: 79 (1968). —Lock in F.T.E.A., Cannaceae: 7 (1993). Type: Not decided, one of two specimens in Herb. Royen (L photo). FIGURE 13.4.**25**.
  Canna edulis Ker Gawl. in Bot. Reg. **9**: t.775 (1823). Type: Plate 775 in Bot. Reg., drawn from a plant grown in England from seed from Peru.
  Canna orientalis Roscoe, Monandr. Pl.: t.12 (1826). Type: Roscoe's plate 12, drawn from a plant of unknown origin grown in Liverpool, England.
  Canna bidentata Bertol. in Mem. Acad. Sci. Bologna **10**: 33, t.5 (1859). Type: Plant cultivated at Bologna from seed from Mozambique (BO).
  Canna indica subsp. *orientalis* (Roscoe) Baker in F.T.A. **7**: 328 (1898).

Perennial erect herb with 1–many stems up to 1–1.5(2) m high from a horizontal rhizome. Leaves sessile, lamina 15–33(60) × 8–16(25) cm, narrowly to broadly ovate to elliptic, acute to acuminate at the apex, rounded to cuneate at base, tapering to the sheath, with very numerous, closely spaced lateral nerves, slightly curved to the apex, not prominent; margins entire. Inflorescence racemose, terminal, pedunculate, usually simple, sometimes branched progressively later, bearing single or paired flowers. Bracts 0.9–2(2.5) × 1(1.5) cm, obovate. Pedicels up to 5 mm long. Sepals 1–1.5 × 0.4–0.9 cm, ovate, acute. Corolla 4–6 cm long, usually red, sometimes flecked with yellow spots, the segments united into a tube from the

Fig. 13.4.**25**. CANNA INDICA. 1, flowering shoot (× ²/₃); 2, leaf (× ²/₃); 3, fruits (× 1). 1–2 from *Cadet* 5839, 3 from *Bos* 1234. Drawn by Christine Grey-Wilson. From Flora des Mascareignes.

base up to 1cm high, lobes free; lobes 3–5 × 0.3–0.8 cm, narrowly elliptic, acute at apex. Staminodes (2)3, reddish, sometimes unequal, 4–6 × 1–1.5 cm, spathulate, acute or strongly emarginate at apex, united at base. Labellum 4–5.5 × 0.6–1 cm, narrowly oblong, emarginate at apex, usually reddish above, yellow with red spots below. Stamen 4–5 cm long, petaloid portion involute; anther 0.7–1 cm long, adnate to sterile portion for a third of its length. Style 4–5 cm long, red or yellow. Capsule 1.5–3.5 × 1.3–2 cm, ovoid with a softly spinose outer layer that falls off before the rest of the capsule opens longitudinally but irregularly. Seeds several, 0.5 cm in diameter, globose, dark brown to blackish, smooth, very hard.

Zambia. C: Luangwa Valley, Musazya Nsefu, cult., fl.& imm.fr. 5.iv.1968, *Phiri* 130 (K). E: Chipata Dist., Chipata (Fort Jameson), 1050 m, fl.& fr. 6.vi.1954, *Robinson* 835 (K). **Zimbabwe**. N: Mazowe, Iron Mask Hill, 1500 m, fl. iii.1906, *Eyles* 253 (BM, K, SRGH). W: Matobo Dist., fr. ii.1954, *Miller* 2225 (SRGH). C: Marondera Dist., site of old Mashona refuge from Matabele, 1350 m, fr. 5.iv.1950, *Wild* 3322 (K, LISC, SRGH). E: Chipinge Dist., near Chirinda, 1140 m, fl. iii.1906, *Swynnerton* 398 (BM). S: Masvingo Dist., Great Zimbabwe (Zimbabwe Ruins), fl. 1.vii.1930, *Hutchinson & Gillett* 3329 (K). **Malawi**. N: Chitipa Dist., Misuku, 1500 m, streambank, fl. 14.xi.1952, *Williamson* 110 (BM). S: Mulanje Dist., Kundwelo village, Palombe plain, 630 m, fl.& fr. 29.vii.1956, *Newman & Whitmore* 282 (BM, SRGH). **Mozambique**. Z: Quelimane, fr. 14.x.1965, *Mogg* 32438 (LISC, SRGH). MS: Sussundenga Dist., Mavita, between Moribane and school, by R. Chicure, fl. 13.i.1948, *Barbosa* 828 (BR, COI, LISC). GI: Chibuto Dist., Chipenhe, c.10 km towards Mainguelane, Chiridzene Forest (Mata de Chirindzene), fr. 19.ix.1980, *Nuvunga, Boane & Conjo* 337 (K, LMU).

Native of South America but now with a pantropical distribution. On stream and dam banks, among ruins and on granite rocky places; 0–1500 m.

Conservation notes: A widespread species, almost certainly introduced and naturalised in the Flora area; not threatened.

This species and various hybrids are cultivated ("Indian Shot") in gardens for their flowers.

MARANTACEAE

by J.M. Lock & M.A. Diniz

Perennial herbs. Stems hard and stiff but without secondary thickening. Leaves in two rows, made up of sheath, false petiole and blade, the sheath open. Lamina with a prominent midrib and close parallel lateral nerves diverging obliquely from midrib. Flowers hermaphrodite, asymmetric, in terminal bracteate spike, raceme or panicle, or inflorescence arising from rhizome. Perianth usually differentiated into calyx and corolla; outer segments free, inner ones fused into a tube, at least at base, unequal. Androecium composed of one fertile stamen with single fertile anther-cell, and several variously petaloid staminodes. Ovary inferior, 1–3-celled; style stout, often apically dilated. Fruit fleshy, or a capsule. Seeds with abundant endosperm, often arillate.

About 30 genera and 550 species, mainly in the wet tropics, most abundant and diverse in the New World. The starchy roots of *Maranta arundinacea* L. are the source of the 'arrowroot' of commerce, which is used as a thickening agent in cooking. Species of *Maranta* and *Calathea* are cultivated as ornamentals, and the aril of the West African *Thaumatococcus daniellii* Benth. is intensely sweet and has been investigated as a possible sugar substitute.

Two species of Marantaceae are recorded by Maroyi (Kirkia **18**: 187, 2006) as having been grown in Zimbabwe: *Calathea ornata* (Lindl.) Korn., a perennial ornamental herb (C: *Biegel* 3231, SRGH), and *Maranta arundinacea* L., a perennial herb cultivated for its edible tubers (C: *Whellan* s.n., SRGH).

THALIA L.

Thalia L., Sp. Pl.: 1193 (1753).

Perennial herbs. Basal leaves petiolate. Inflorescence laxly paniculate, much-branched with slender zigzag branchlets, each one with an adjacent leaf-like bract. Flowers in pairs, enclosed in a pair of bracts. Sepals 3, equal, membranaceous. Petals 3, free or slightly joined at base, posterior rather larger than the others. Staminodes 3, petaloid, in 2 whorls. Ovary 1-locular, with 1 ovule, globose. Fruit capsular, subglobose to ellipsoid, 1-seeded. Seed oblong-ellipsoid, arillate.

Around 7 species in tropical America, one extending to Africa.

Thalia geniculata L., Sp. Pl.: 1193 (1753). —Baker in F.T.A. **7**: 314 (1898). — Andersson in Nordic J. Bot. **1**: 48 (1981). Type: t.58, fig.1 in Plumier's Plant. Americanum (ed. J. Burmannus) (1755). FIGURE 13.4.**26**.

  Thalia caerulea Ridl. in J. Bot. **25**: 132 (1887). —Baker in F.T.A. **7**: 314 (1898). —Rendle in Cat. Afr. Pl. Welw. **2**: 22 (1899). Type: Angola, Pungo Andongo, marshes by R. Cuanza, Sobati Nbille, *Welwitsch* 6444 (BM holotype, LISU).

  Thalia welwitschii Ridl. in J. Bot. **25**: 132 (1887). —Baker in F.T.A. **7**: 314 (1898). — Rendle in Cat. Afr. Pl. Welw. **2**: 22 (1899). —Hepper in F.W.T.A. ed.2, **3**: 85 (1968). Types: Angola, Cazengo, between Cacula and Dalatando, *Welwitsch* 6443 (BM syntype, LISU) & Angola, Pungo Andongo, Mutollo, *Welwitsch* 6445 (BM syntype, LISU).

Perennial rhizomatous herb 1–3 m tall. Basal leaves petiolate, stem leaves epetiolate; pulvinus 0.3–2.5 cm long; lamina 19–63 × 4–26 cm, usually ovate to elliptic or narrowly so, gradually and shortly acuminate at apex, rounded or subcordate at base, glabrous, often glaucous beneath. Inflorescence much-branched, laxly paniculate with slender branches; at first with overlapping glabrous to densely hirsute spathes, not persisting; rachis flexuose, glabrous to hirsute. Spathes 11–22 mm long, usually glabrous. Sepals 3, 2–3 mm long, scale-like, glabrous to densely hirsute, hyaline or greenish with a scarious margin. Petals pale mauve, 3, 6–13 mm long, posterior one often only half as long as laterals, glabrous to sparsely pilose. Staminodes 3, the outer petaloid, 15–20 × 5–10 mm, mauve, others largely firm and fleshy with narrow petaloid rim, one bearing a single fertile theca. Ovary 1-locular, 1-ovular, glabrous or sparsely pilose. Fruit ellipsoid to subglobose, up to 10 mm long, indehiscent; pericarp very thin, drying papery. Seed filling the fruit, broadly oblong-ellipsoid, smooth, brownish to greyish or black, with a basal aril.

Zambia. B: Mongu Dist., Bulozi Plain, below Mongu, fl. 9.xi.1959, *Drummond & Cookson* 6268 (LISC, SRGH). N: Mansa Dist., Mansa (Fort Rosebery), from side of Goodall channel c.153 km from Lake Chali, fl. 17.ii.1959, *Watmough* 273 (K, LISC, SRGH). C: Kabwe Dist., Mwapula R., Mwiya village, 14.v.1995, *Bingham* 10533 (K).

In Africa found in West Africa from Senegal to S Sudan; also in Angola and Congo. Elsewhere found in the New World from Florida in the north to tropical Argentina in the south, but absent from much of the Amazon Basin. Andersson (1981) lists Zimbabwe in his summary of the distribution, but cites no specimens; we have not seen any. Swamp vegetation by lakes, rivers and channels; sometimes in roadside ditches; c.1000 m.

Conservation notes: A widespread species found in habitats that are extensive within its range; not threatened.

Thalia welwitschii has been treated as a distinct African species, but Andersson has shown by careful morphometric analysis that the African and South American plants are part of a single complex for which *T. geniculata* is the earliest name. He points out, however, that most African plants have a reduced posterior petal – a character that is also found in some plants in the New World, particularly in the Greater Antilles – and an aril with the ends evolute rather than involute. He did not find any

Fig. 13.4.**26**. THALIA GENICULATA. 1, leaf and flowering shoot; 2, flower; 3, androecium; 4, style; 5, fruit with part of pericarp removed; 6 seed. Source specimen and sizes not known. Drawn by Stella Ross-Craig. From Flora of West Tropical Africa.

American plants showing this latter feature. Most of the specimens at Kew from the Flora area have narrower leaves than those of Neotropical plants. Andersson also pointed out that the range of variation within Africa is much less than that in America, and suggests that this is the result of a relatively recent introduction of a small number of founder individuals. However, the species occurs in both northern and southern tropical Africa, with a very marked disjunction between the two regions – the species has not been recorded from the F.T.E.A. area – and occurs in apparently natural habitats. If the species reached Africa relatively recently, this would seem more likely to have been through natural dispersal.

For details of typification and full synonymy see Andersson (1981).

INDEX TO BOTANICAL NAMES

FAMILIES OF VASCULAR PLANTS REPRESENTED IN THE FLORA ZAMBESIACA AREA

PTERIDOPHYTA
(Flora Zambesiaca families and family number. Published 1970)

Actiniopteridaceae		Gleicheniaceae	9	Parkeriaceae	
see Adiantaceae	18	Grammitidaceae	20	see Adiantaceae	18
Adiantaceae	18	Hymenophyllaceae	15	Polypodiaceae	21
Aspidiaceae	27	Isoetaceae	4	Psilotaceae	1
Aspleniaceae	23	Lindsaeaceae	19	Pteridaceae	
Athyriaceae	25	Lomariopsidaceae	26	see Adiantaceae	18
Azollaceae	13	Lycopodiaceae	2	Salviniaceae	12
Blechnaceae	28	Marattiaceae	7	Schizaeaceae	10
Cyatheaceae	14	Marsileaceae	11	Selaginellaceae	3
Davalliaceae	22	Oleandraceae		Thelypteridaceae	24
Dennstaedtiaceae	16	see Davalliaceae	22	Vittariaceae	17
Dryopteridaceae		Ophioglossaceae	6	Woodsiaceae	
see Aspidiaceae	27	Osmundaceae	8	see Athyriaceae	25
Equisetaceae	5				

GYMNOSPERMAE
(Flora Zambesiaca families and family number. Volume 1(1) 1960)

Cupressaceae	3	Cycadaceae	1	Podocarpaceae	2

ANGIOSPERMAE
(Flora Zambesiaca families, volume and part number and year of publication)

Acanthaceae	–	–	Balsaminaceae	2(1)	1963
Agapanthaceae	13(1)	2008	Barringtoniaceae	4	1978
Agavaceae	13(1)	2008	Basellaceae	9(1)	1988
Aizoaceae	4	1978	Begoniaceae	4	1978
Alangiaceae	4	1978	Behniaceae	13(1)	2008
Alismataceae	12(2)	2009	Berberidaceae	1(1)	1960
Alliaceae	13(1)	2008	Bignoniaceae	8(3)	1988
Aloaceae	12(3)	2001	Bixaceae	1(1)	1960
Amaranthaceae	9(1)	1988	Bombacaceae	1(2)	1961
Amaryllidaceae	13(1)	2008	Boraginaceae	7(4)	1990
Anacardiaceae	2(2)	1966	Brexiaceae	4	1978
Anisophylleaceae			Bromeliaceae	13(2)	2010
see Rhizophoraceae	4	1978	Buddlejaceae		
Annonaceae	1(1)	1960	see Loganiaceae	7(1)	1983
Anthericaceae	13(1)	2008	Burmanniaceae	12(2)	2009
Apocynaceae	7(2)	1985	Burseraceae	2(1)	1963
Aponogetonaceae	12(2)	2009	Buxaceae	9(3)	2006
Aquifoliaceae	2(2)	1966	Cabombaceae	1(1)	1960
Araceae	–	–	Cactaceae	4	1978
Araliaceae	4	1978	Caesalpinioideae		
Aristolochiaceae	9(2)	1997	see Leguminosae	3(2)	2006
Asclepiadaceae	–	–	Campanulaceae	7(1)	1983
Asparagaceae	13(1)	2008	Canellaceae	7(4)	1990
Asphodelaceae	12(3)	2001	Cannabaceae	9(6)	1991
Avicenniaceae	8(7)	2005	Cannaceae	13(4)	2010
Balanitaceae	2(1)	1963	Capparaceae	1(1)	1960
Balanophoraceae	9(3)	2006	Caricaceae	4	1978

Family	Vol(Part)	Year
Caryophyllaceae	1(2)	1961
Casuarinaceae	9(6)	1991
Cecropiaceae	9(6)	1991
Celastraceae	2(2)	1966
Ceratophyllaceae	9(6)	1991
Chenopodiaceae	9(1)	1988
Chrysobalanaceae	4	1978
Colchicaceae	12(2)	2009
Combretaceae	4	1978
Commelinaceae	–	–
Compositae		
tribes 1–5	6(1)	1992
Connaraceae	2(2)	1966
Convolvulaceae	8(1)	1987
Cornaceae	4	1978
Costaceae	13(4)	2010
Crassulaceae	7(1)	1983
Cruciferae	1(1)	1960
Cucurbitaceae	4	1978
Cuscutaceae	8(1)	1987
Cymodoceaceae	12(2)	2009
Cyperaceae	–	–
Dichapetalaceae	2(1)	1963
Dilleniaceae	1(1)	1960
Dioscoreaceae	12(2)	2009
Dipsacaceae	7(1)	1983
Dipterocarpaceae	1(2)	1961
Droseraceae	4	1978
Ebenaceae	7(1)	1983
Elatinaceae	1(2)	1961
Ericaceae	7(1)	1983
Eriocaulaceae	13(4)	2010
Eriospermaceae	13(2)	2010
Erythroxylaceae	2(1)	1963
Escalloniaceae	7(1)	1983
Euphorbiaceae	9(4)	1996
Euphorbiaceae	9(5)	2001
Flacourtiaceae	1(1)	1960
Flagellariaceae	13(4)	2010
Fumariaceae	1(1)	1960
Gentianaceae	7(4)	1990
Geraniaceae	2(1)	1963
Gesneriaceae	8(3)	1988
Gisekiaceae		
see Molluginaceae	4	1978
Goodeniaceae	7(1)	1983
Gramineae		
tribes 1–18	10(1)	1971
tribes 19–22	10(2)	1999
tribes 24–26	10(3)	1989
tribe 27	10(4)	2002
Guttiferae	1(2)	1961
Haloragaceae	4	1978
Hamamelidaceae	4	1978
Hemerocallidaceae	12(3)	2001
Hernandiaceae	9(2)	1997
Heteropyxidaceae	4	1978
Hyacinthaceae	9(2)	1997
Hydnoraceae	9(2)	1997
Hydrocharitaceae	12(2)	2009
Hydrophyllaceae	7(4)	1990
Hydrostachyaceae	9(2)	1997
Hypericaceae		
see Guttiferae	1(2)	1961
Hypoxidaceae	12(3)	2001
Icacinaceae	2(1)	1963
Illecebraceae	1(2)	1961
Iridaceae	12(4)	1993
Irvingiaceae	2(1)	1963
Ixonanthaceae	2(1)	1963
Juncaceae	13(4)	2010
Juncaginaceae	12(2)	2009
Labiatae		
see Lamiaceae, Verbenacaeae		
Lamiaceae		
Viticoideae, Pingoideae	8(7)	2005
Lamiaceae		
Scutellaroideae-		
Nepetoideae	–	–
Lauraceae	9(2)	1997
Lecythidaceae		
see Barringtoniaceae	4	1978
Leeaceae	2(2)	1966
Leguminosae,		
Caesalpinioideae	3(2)	2007
Mimosoideae	3(1)	1970
Papilionoideae	3(3)	2007
Papilionoideae	3(4)	–
Papilionoideae	3(5)	2001
Papilionoideae	3(6)	2000
Papilionoideae	3(7)	2002
Lemnaceae	–	–
Lentibulariaceae	8(3)	1988
Liliaceae sensu stricto	12(2)	2009
Limnocharitaceae	12(2)	2009
Linaceae	2(1)	1963
Lobeliaceae	7(1)	1983
Loganiaceae	7(1)	1983
Loranthaceae	9(3)	2006
Lythraceae	4	1978
Malpighiaceae	2(1)	1963
Malvaceae	1(2)	1961
Marantaceae	13(4)	2010
Mayacaceae	13(2)	2010
Melastomataceae	4	1978
Meliaceae	2(1)	1963
Melianthaceae	2(2)	1966
Menispermaceae	1(1)	1960
Menyanthaceae	7(4)	1990
Mesembryanthemaceae	4	1978
Mimosoideae		
see Leguminosae	3(1)	1970
Molluginaceae	4	1978
Monimiaceae	9(2)	1997
Montiniaceae	4	1978
Moraceae	9(6)	1991
Musaceae	13(4)	2010

Myristicaceae	9(2)	1997	Rhamnaceae	2(2)	1966
Myricaceae	9(3)	2006	Rhizophoraceae	4	1978
Myrothamnaceae	4	1978	Rosaceae	4	1978
Myrsinaceae	7(1)	1983	Rubiaceae		
Myrtaceae	4	1978	subfam. Rubioideae	5(1)	1989
Najadaceae	12(2)	2009	tribe Vanguerieae	5(2)	1998
Nesogenaceae	8(7)	2005	subfam.Cinchonoideae	5(3)	2003
Nyctaginaceae	9(1)	1988	Rutaceae	2(1)	1963
Nymphaeaceae	1(1)	1960	Salicaceae	9(6)	1991
Ochnaceae	2(1)	1963	Salvadoraceae	7(1)	1983
Olacaceae	2(1)	1963	Santalaceae	9(3)	2006
Oleaceae	7(1)	1983	Sapindaceae	2(2)	1966
Oliniaceae	4	1978	Sapotaceae	7(1)	1983
Onagraceae	4	1978	Scrophulariaceae	8(2)	1990
Opiliaceae	2(1)	1963	Selaginaceae		
Orchidaceae	11(1)	1995	see Scrophulariaceae	8(2)	1990
Orchidaceae	11(2)	1998	Simaroubaceae	2(1)	1963
Orobanchaceae			Smilacaceae	12(2)	2009
see Scrophulariaceae	8(2)	1990	Solanaceae	8(4)	2005
Oxalidaceae	2(1)	1963	Sonneratiaceae	4	1978
Palmae	13(2)	2010	Sphenocleaceae	7(1)	1983
Pandanaceae	12(2)	2009	Sterculiaceae	1(2)	1961
Papaveraceae	1(1)	1960	Strelitziaceae	13(4)	2010
Papilionoideae			Taccaceae		
see Leguminosae	–	–	see Dioscoreaceae	12(2)	2009
Passifloraceae	4	1978	Tecophilaeaceae	12(3)	2001
Pedaliaceae	8(3)	1988	Tetragoniaceae	4	1978
Periplocaceae			Theaceae	1(2)	1961
see Asclepiadaceae	–	–	Thymelaeaceae	9(3)	2006
Philesiaceae			Tiliaceae	2(1)	1963
see Behniaceae	13(1)	2008	Trapaceae	4	1978
Phormiaceae			Turneraceae	4	1978
see Hemerocallidaceae	12(3)	2001	Typhaceae	13(4)	2010
Phytolaccaceae	9(1)	1988	Ulmaceae	9(6)	1991
Piperaceae	9(2)	1997	Umbelliferae	4	1978
Pittosporaceae	1(1)	1960	Urticaceae	9(6)	1991
Plantaginaceae	9(1)	1988	Vacciniaceae		
Plumbaginaceae	7(1)	1983	see Ericaceae	7(1)	1983
Podostemaceae	9(2)	1997	Vahliaceae	4	1978
Polygalaceae	1(1)	1960	Valerianaceae	7(1)	1983
Polygonaceae	9(3)	2006	Velloziaceae	12(2)	2009
Pontederiaceae	13(2)	2010	Verbenaceae	8(7)	2005
Portulacaceae	1(2)	1961	Violaceae	1(1)	1960
Potamogetonaceae	12(2)	2009	Viscaceae	9(3)	2006
Primulaceae	7(1)	1983	Vitaceae	2(2)	1966
Proteaceae	9(3)	2006	Xyridaceae	13(4)	2010
Ptaeroxylaceae	2(2)	1966	Zannichelliaceae	12(2)	2009
Rafflesiaceae	9(2)	1997	Zingiberaceae	13(4)	2010
Ranunculaceae	1(1)	1960	Zosteraceae	12(2)	2009
Resedaceae	1(1)	1960	Zygophyllaceae	2(1)	1963
Restionaceae	13(4)	2010			